To the most amazing mother-in-law, Glenda Cherry.

From day dot you saw my tireless love for gardening
as an extension of who I am. You simply saw me, and
for this I could not imagine a life without you.

The
Plant Society
Design Handbook

JASON CHONGUE

murdoch books

Sydney | London

Contents

Introduction

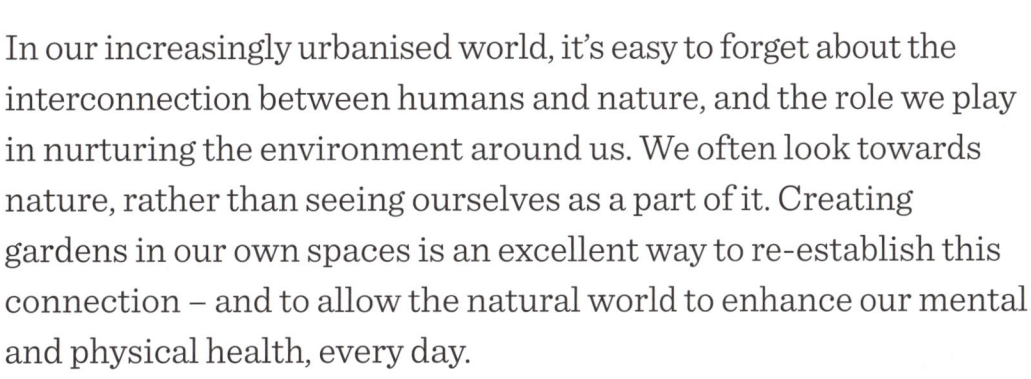

In our increasingly urbanised world, it's easy to forget about the interconnection between humans and nature, and the role we play in nurturing the environment around us. We often look towards nature, rather than seeing ourselves as a part of it. Creating gardens in our own spaces is an excellent way to re-establish this connection – and to allow the natural world to enhance our mental and physical health, every day.

I've always felt complete when in the garden; my relationship with plants has been a solace and source of calm. I started The Plant Society in 2016 with my partner, Nathan, with the simple goal of introducing that joy into people's lives.

At the time, I was working as an architect with a focus on interiors, and I was passionate about all the ways we interact with design and move through our homes and workplaces. Design can of course be replicated, but it's the individual, personal touches that make a space feel like a home – and which set it apart. For me, greenery is key to this, the comfort and completeness that plants provide being a crucial element in the success of a design. We tend to think a lot about how our homes look and feel, paying particular attention to architectural and interior design elements. While the conversation has changed since 2016, plants – and the budget for them! – are still often considered a nice-to-have or an afterthought. Yet they are about more than growing a garden: they allow you to curate a space that breathes your heart and soul.

We launched our business from this perspective – not as horticulturalists, but as designers with a fresh take on plants and gardening. We provide support to people who want to bring plants into their life, whether it's one pot or a whole garden, whether it's in a bathroom, across multiple rooms, on a verandah, or outside. We strip gardening back to basics so that nurturing and maintaining plants feels achievable for everyone. We have curated gardens for clients across Australia, from pocket-sized green spaces in apartments to large public installations and corporate spaces. My approach to a successful garden design always starts with two simple things: functionality and personality. The green elements in your space must practically suit your needs and lifestyle, plus they should reflect your own tastes and values.

Many people still think of gardens as exclusively green spaces outside. But if you ask me, a garden is any space where you can tend

11

to greenery. Anything from a cluster of three indoor plants to a large backyard with hundreds of plants – and even a shared verge planting in a community. There really are no rules about what form a garden can or should take.

'But I'm not a gardener', I hear you say. Well, neither was I. My love for plants grew out of a happy childhood hobby. I didn't train in horticulture; rather, I've built on the knowledge that fellow gardeners have shared with me over more than 25 years. My approach to gardening is a reflection of who I am: I combine intuition with experimentation and, mostly, get where I want to go. One of my greatest satisfactions is sharing what I've learned, giving others the skills and confidence to get to know the rewards of their own personal green spaces.

WHY THIS BOOK?

I believe that we can all create and maintain a well-designed green space for our home, no matter what our budget or lifestyle. With this handbook, I hope to inspire you to dream up your own green space, big or small, and give you all of the ingredients you need to bring this vision to life.

My goal is to talk you through the design fundamentals to consider when planning your space, and to share the knowledge I've gleaned from working with customers and clients, incorporating the core design elements: colour, form, texture, layout and scale. Simple theories, such as the art of grouping (see page 66), can transform randomly placed plants into a room or space's central talking point – and highlight.

While there is no shortage of gardening books, including those that focus on indoor plants, what compelled me to write this, my third book, is the absence of what I call 'recipes' and blueprints. I have always enjoyed being inspired by dishes that chefs create and pass on through their recipes. That was my inspiration here – to promote the sharing of knowledge in a straightforward way and to adapt the recipe concept to make gardening and designing gardens more accessible to more people.

The 41 recipes and blueprints include illustrations, photography, plan sketches where applicable, a Kit of Parts much like an IKEA installation guide, a shopping list (including plant types, planter shapes and potting mix types), and tips for ongoing care. I've gone to town (limited only by the number of pages in this book), and examples include potted cluster gardens; a tiny terrace with herbs and citrus; a bold balcony arrangement; a lush, narrow fence line;

an indoor winter garden; a verandah hanging garden; verdant plant shelves and a green workspace.

These easy-to-follow recipes and blueprints provide designs for small gardens that can be created in a day, or over a weekend, and include photos and illustrations to guide and, I hope, inspire you. In each design, I've aimed to create an atmosphere specific to the room or space. But if you fall in love with the design and want to use it somewhere else, then go for it! As far as I'm concerned, if I've inspired you on that path, then my work is done.

Of course, it wouldn't be a useful garden handbook without the basics. So, they're all here, too, including information on getting started (it sounds obvious, but it's often the biggest hurdle); budgeting; understanding seasonality; when to plant; when to sow; and the art of plant (and planter) choice (it *is* an art!). And then, once your garden is up and flourishing, essential knowledge for keeping it that way: regular plant care (such as watering); pruning; deadheading; repotting; aerating; feeding; beating pests; and more.

In the creation of this book, I've visited countless homes with photographer Armelle Habib. There's so much inspiration out there, and our hope is that in bringing you a slice of it, we'll help you to feel empowered to go out and create your own green oasis. With one disclaimer: it's addictive. ■

analysing your space

I consider a garden to be any green space – inside or out – and this includes courtyards, rooftops, balconies, shelves, backyards, small potted clusters, and plants in the ground. There are no limits to what you can envisage as your garden.

PLANNING

When planning a garden, too often we rush the preparation process. Gardens don't need to be perfect. As our lives change, so do the things we prefer. So, as your garden evolves, it's also okay to edit and change your path.

Whenever I start the process of planning a garden, I pause and observe the space to collect my thoughts and set up the foundations to keep my ideas on track. Below is a list of things to look for when observing the fundamentals of your space. See page 22 for more detail on some of these elements.

INDOOR

LIGHTING
> Where are the windows located?
> What direction are your windows facing?

FURNITURE & OTHER STYLING OBJECTS
> What interior elements are you working with?

SPACE
> What spatial constraints are present? This includes ceiling height.

EXISTING INTERIOR CONCEPT
> To create a seamless interior, it is good practice to carry the interior language into your plant design.

OUTDOOR

LIGHTING
> What exposure does your garden receive?

ARCHITECTURE
> What existing elements do you need to consider?

SPACE
> What spatial constraints are present?

ACCESS
> Are there areas you need to maintain for access and functionality?

understanding

your space

There are many elements to consider when trying to understand your space, from the amount of sunlight and the environmental conditions to the presence of existing structures.

Lighting

Note the orientation of the space and how the sun moves over it. Whether indoors or out, gardens rely on sunlight – and selecting the right plants for the lighting situation will set you on the path to success. When you understand the intensity and duration of natural lighting on hand, you can choose plants that prefer those conditions. My biggest tip when selecting a plant for your garden is to think about where the plant originated and what lighting conditions it would be exposed to in its natural habitat. This will help you understand the plant's needs.

Wind/drafts

We tend to forget the role that wind plays in our gardens. Wind creates a challenge when it tears through foliage and impacts the growth of your plants. For instance, exposed balconies are open to the elements, so wind can topple plants and rapidly dry out soil. Indoors, heating and cooling vents can remove moisture from the atmosphere. Early planning allows you to manage these issues. Simple measures – such as using pots without curved bases – can minimise the risk of your plants being blown over. Another approach I take in windy spaces is to choose plants that have naturally adapted to the harsh conditions prevalent in desert and Mediterranean climates.

Soil conditions

Soil is an important element in any garden, and different plants need different soil types in order to thrive. However, this doesn't mean that you need a specific soil type for each plant. Instead, you can find a happy medium that will generally satisfy the bulk of your plant selection (see pages 110–11).

Water/rain

Each plant has its own watering requirements. Knowing your local rainfall levels and working out the amount of hand-watering to which you can commit will guide you in your plant choices.

Planting for seasons and climate

Choosing plants that will tolerate the climate and the seasonal changes in your area will make for an easy-care and interesting garden. The amount of sunlight typically changes with the seasons, as does the temperature.

The space at hand

How much physical space do you have? It is a good idea to measure out your space so you can create a garden that is at the right scale.

New and existing architecture

Consider the built structures or forms that may enhance or hinder your planting. For instance, if an apartment building is under construction next door, then this may change the amount of natural lighting that your garden receives. On the other hand, a structure such as a fence could help to protect your plants from extreme winds, providing shelter as they grow. Take note of the surrounding built forms, both existing and planned.

GREEN TIP
ALL PLANTS NEED
SOME LEVEL OF
SUNLIGHT TO THRIVE.
ENSURE THAT THE
AMOUNT AND QUALITY
OF AVAILABLE LIGHT
ARE ADEQUATE FOR
INDOOR PLANTS.

understanding

yourself

We don't often consider who we are as gardeners. Are you happy to spend every moment of daylight tending to your garden, or just one hour per week? How much greenery can you care for? Do you have a busy work and social life? These questions are important to consider when creating a garden, and the answers will likely change throughout your life.

As a passionate gardener, I tend to spend hours on end in my indoor and outdoor green spaces. I understand that this won't be everybody's choice, and that's fine – we all find joy from gardening at different levels.

Think about how much time you would like to spend looking after your garden. This will assist you in curating a space that reflects the level of attention you can offer. I do want to stress, though, that every garden needs *some* nurturing, even if it's only minimal.

If you're only going to be spending a small amount of time in the garden, think about using low-maintenance plants that will thrive in the conditions at hand. However, if you want to really immerse yourself in your garden, then try species that will challenge your skill set.

understanding your style

What kind of garden do you prefer? Lush and wild, or neat and ordered? Having a good handle on your personal style will help you during both the garden-design and plant-selection processes.

Look and Feel document

One of the best ways to start your design process is to create a Look and Feel document. Consider it a refined scrapbook. How you put together this scrapbook is entirely up to you. I like to create mine digitally: I lay out all of my materials and inspirations side by side in a document. The images come from websites or social media, or they are photos I've taken and saved over the years. If you prefer to create a physical scrapbook, then this is also perfectly fine.

This is a great way to seamlessly catalogue your ideas and to create a clear visual guide for the style you'd like to produce and the emotions you want your space to elicit. Add all of the elements you love and wish to use in your design. Then look at the document with a critical eye: edit out any images and ideas that don't fit into the desired style or feel. It can be difficult to get rid of things you adore, but it's better to do it now, during the planning period.

The height of the tall plant is mirrored by the lamp. This adds balance to the space.

Your Look and Feel document will help to keep your ideas consistent during the entire garden-design process. (For more information on how to put this into practice to create your garden, see page 30.)

Creating atmosphere

We often think about the spaces in our home as separate from one another. We see gardens as being external, isolated from furniture and internal design.

I like to consider all elements of a space holistically and to create an atmosphere through attention to detail. Simple decisions – such as using similar tones in both your soft furnishings and your planters – can make for a stronger aesthetic.

How much greenery?

There is no right or wrong answer to the question of how much greenery will make a space cosy. Instead, it's a question of how much maintenance you're comfortable with, and what level of gardening you can balance with your lifestyle. Being a gardener should be enjoyable and not feel like a chore.

As I get older, my relationship with gardening is evolving. My time in the garden allows me to reflect on who I am. Sometimes, I ponder big life questions. Other times, I am just in a state of bliss as I enjoy my surroundings. Sometimes, I want the level of cultivation to be minimal, while during other periods I'm happy to do more in the garden. Whether you're an avid or casual gardener, your relationship with gardening will change – so, it's important to keep this relationship fluid.

Your Look and Feel document will also help you to decide on the right amount of greenery for your garden. Rather than placing plants all over your space, I recommend that you concentrate your planting into considered areas to create visual 'moments' that reflect your chosen design style. This will direct the focus to key areas rather than forcing the eye to dart around.

If you aren't quite sure how many plants you need, then start small and gradually add plants. On the other hand, if you're a gardening fanatic, then don't suppress that urge to plant more.

ANY GREAT SPACE
HAS ATMOSPHERE. YOUR
LOOK AND FEEL DOCUMENT
SHOULD INCLUDE THE
SPECIFIC MOOD YOU WANT
TO CREATE, SO YOU
CAN PLAN IT INTO
THE SPACE.

KIT OF PARTS

When starting a design for any space, I begin with a simple, personal guidebook: my 'Kit of Parts'. This is a handy tool that helps you create and stay within defined design parameters. Putting together your own Kit of Parts will help to inform many decisions you'll make on your garden-design journey.

You can create your Kit of Parts digitally or as a physical document, whichever is most comfortable for you. It's important that you can easily reference this document when you're curating, sourcing and installing your garden project, as it will ensure that you stay aligned with your concept and make considered design decisions.

Your Kit of Parts is based on your Look and Feel document (see pages 25–6). Once you've edited out the anomalies in your Look and Feel document, cut it back again until you're left with a few images and ideas that really speak to one another – I recommend restricting yourself to three images and/or ideas to make sure that you've made all of the hard decisions up-front. Then you can add to your Kit of Parts all of the items you need to physically create your design. See page 32 for the elements to consider when choosing items for your Kit of Parts.

Building your Kit

Short-list the planters and plants that will be used in the project, and add them to your Kit of Parts. (For more details on plant types and suitable conditions, see pages 54–65.) Take cues from your Look and Feel document, but give yourself permission to evolve the concept slightly. After all, you don't want to mimic the design references. When working on potted gardens, start with the planters you want to use, then move on to the plants that will work in the project. If your design doesn't require planters, then move straight into short-listing the plants you want to incorporate.

Remember to refer back to the fundamentals of your space (see page 20). If the plant you love doesn't like the conditions, then it's never going to thrive. I recommend listing all of the plants you love and then researching the conditions each plant needs in order to establish if your selections will flourish in your space.

Finally, think about any additional pieces that you want to include in your garden design. This might be a plant stand, stake, sculpture or even table setting. Add these selections to another page in your Kit of Parts document. Of course, there will also be practical items – such as potting materials, compost, gravel and mulch – that are important to include in your Kit of Parts.

When you're happy with your Kit of Parts, it's time to sign off on it. Try not to change it or add to it during the design process (see Chapter 3), so you stay true to your original design concept.

Elements to consider

Colour

> What colour scheme do you want to embrace? See pages 34–5 for some colour palette examples.
> Will you stick to solid tones or shades?
> Colour could be applied through paints and objects, or it can be added through foliage or flowers.
> Designing with colours that complement each other or purposefully clash can build a strong personality in your designs. Determining these colours prior to selecting plants for your garden helps to focus your concept (see pages 82–7 for more on colour).

Texture

> I think of plant foliage as textural, like fabric. You could break up texture into categories – such as chunky, fine, glossy and matt – to help you refine your choices. See page 52 for a list of interesting foliage textures.
> Layering of foliage (see page 98) can add visual depth to your garden. Have a play with a range of foliage forms, and consider how they may create a push and pull between each other. For instance, having a delicate grass beside the chunky foliage of an agave will create greater definition of each foliage type.
> Texture can also be embraced in your garden through flowers, seed heads, flower stems, fallen leaf material and even the shadows that a tree canopy casts on the ground beneath. As the garden changes throughout the seasons, so will the textures on hand. I adore when lush green grasses such as feather reed grass (*Calamagrostis × acutiflora*) become delicate mounds with downy dried flower heads; they add a softer touch in winter when the garden is predominantly dormant. Bark on tree trunks can also bring interesting textures to your garden. I especially love when the season shifts in autumn, exposing the beautiful trunks of deciduous trees.

Scale

> How tall or wide will a plant grow? Knowing this will allow you to plan for the future. For instance, in a new suburban garden, it might be important to plant trees that will grow tall and command the space. This planning is required to ensure that your concept makes sense for years to come. I usually categorise plants according to height (tall, medium or short).

Growth habits

> The way a plant grows can help to add atmosphere to your space. Consider your space as the canvas, and think about what growth habit will best bring life to your project. To do this, think about the plant's mature form (it helps to imagine the plant's silhouette to keep this task manageable). Consider if the plant will grow tall, stay compact or be a low sprawler. Is it best if it grows upright, or cascades over an edge? See page 52 for more detail on typical growth habits.
> It's important to build a relationship between the scale of your plants and the space they will inhabit. For example, if your space is narrow, then you'll need to use plants that are tall and slender so that the ground level remains functional. If your balcony needs privacy, then you would want to incorporate a dense-growing shrub. There are myriad unique elements that make your space one of a kind, and each space requires a plant with a different growth habit to accommodate onsite characteristics.

A common thread

> When finetuning your Kit of Parts, the best way to edit out ideas that do not fit into the overall picture is to make sure that what is left has a common thread. To do this, lay out the images on a sheet – either digitally or physically – so you can see what goes together and what should be taken away.

Colour inspiration

EARTHY

Inspired by the rich hues found in nature, this popular palette provides a warm, organic feel.

DARK CHOCOLATE

DUSTY ROSE

CARAMEL

COOL

Blues and greens are refreshing and calm in any space. The contrasting splash of red is eye-catching.

FRENCH BLUE

STORM GREY

SAGE

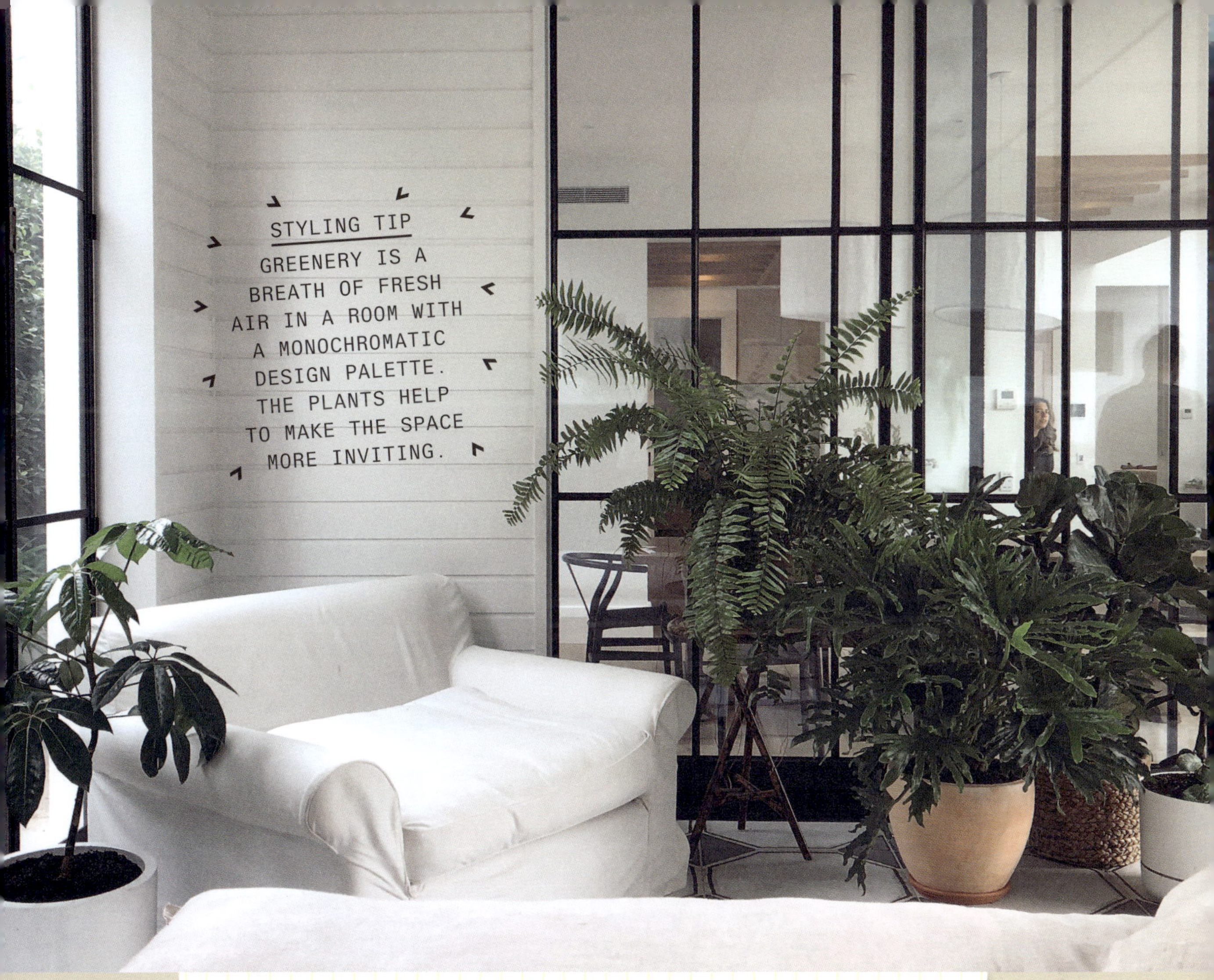
STYLING TIP
GREENERY IS A BREATH OF FRESH AIR IN A ROOM WITH A MONOCHROMATIC DESIGN PALETTE. THE PLANTS HELP TO MAKE THE SPACE MORE INVITING.

MONOCHROMATIC

For a minimalist look, keep to blacks, whites and greys. This allows the greenery to pop and become the focus of attention.

BLACK

WHITE
WHITE LINEN

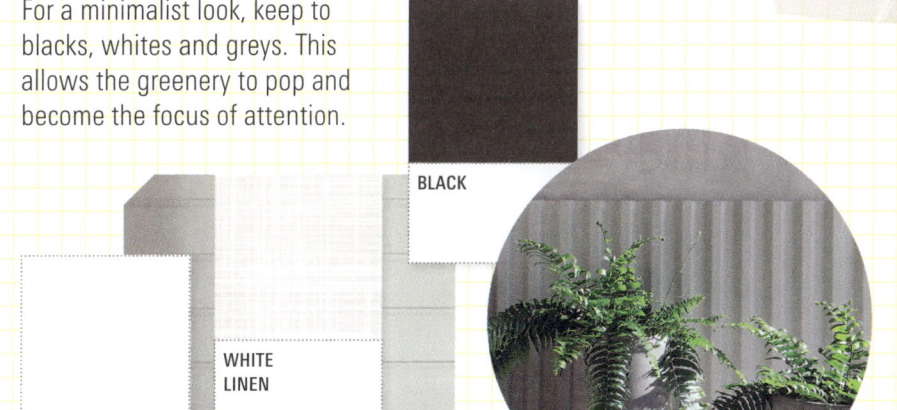

GARDEN LAYOUTS

In my work as a designer and plant curator, the most common questions I get are 'Where should I place greenery?' and 'How do I best lay out my garden?'.

Deciding on a design layout for a green space can be difficult, but over the years I've learned many little tips and tricks (and made a few mistakes!) that I can share to make the job easier for you.

My approach is that spaces should always be functional, while also being aesthetically pleasing. Gardens and plants should be the last pieces that are added to help thread all of the other elements together.

To make your design process easier, on the following pages I've illustrated a series of typical layouts for indoor and outdoor gardens. Find the layout that relates most to your space, then use it to guide you in where to place your plants or how to lay out your garden.

These suggested layouts are based on fundamental design principles and will help you to make the most of your green space.

FAMILY BATHROOM

01 Take advantage of residual spaces in the bathroom. Medium to large floor plants can soften the area around the bath, while small plants will help to break up big double-basin vanities.

02 Even placing a couple of floor plants beside the shower will bring life to tight spaces.

POWDER ROOM

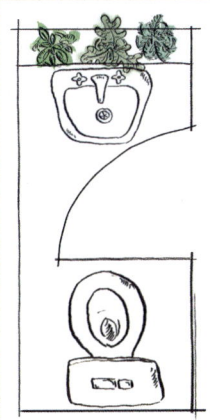 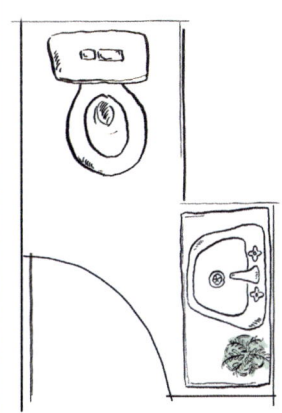

01 Incorporate a mix of smaller plants along the windowsill. Try using a combination of cascading and upright plant types.

02 A single plant beside the basin provides an organic element in any typical powder room.

ENSUITE

 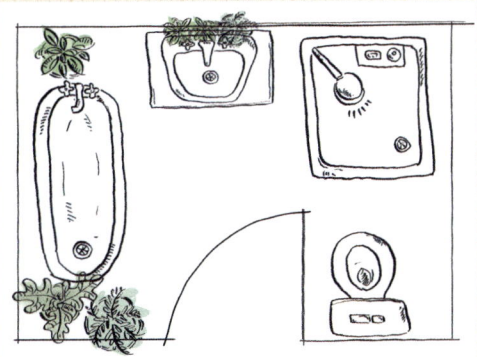

01 Large walk-in showers have enough space for a cluster of plants near the entrance. Try using plants that love humidity, such as rabbit's foot fern (*Davallia solida* var. *fejeensis*).

02 Turn your ensuite into an oasis with a range of different plants placed in corners, both on the floor and above the joinery.

LARGE DINING ROOM

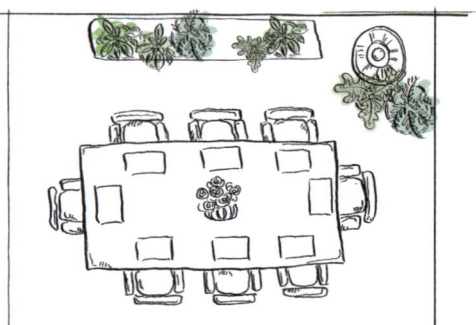

01 Integrate your plant styling with complementary furniture and homewares. This will give you a seamless interior that is perfect for entertaining friends and family.

02 Create simple clusters on your dining table and in an empty corner. Grouping plants in odd numbers helps to provide an organic flow through the room.

APARTMENT DINING ROOM

01 Don't be afraid to create an indoor jungle in a small apartment. To avoid your space feeling chaotic, cluster the plants in large groups.

02 A clever way to create considered focal points in a small space is to limit your planting to just one or two key areas, such as an unused corner.

03 Where floor space is at a premium, look to ledges and furniture for styling opportunities. Cascading plants are perfect for wall shelves.

STUDIO DINING ROOM

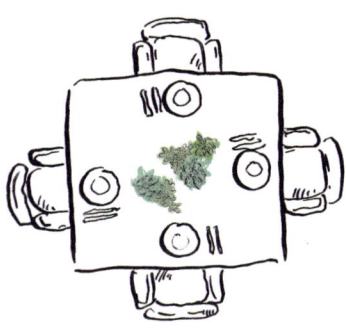

01 If you only have the space or budget for a single plant, try to make it a large plant that will command the room.

02 A cluster of two plants makes for a fresh tabletop feature, especially if they have interesting foliage.

SMALL LIVING ROOM WITH ONE SOFA

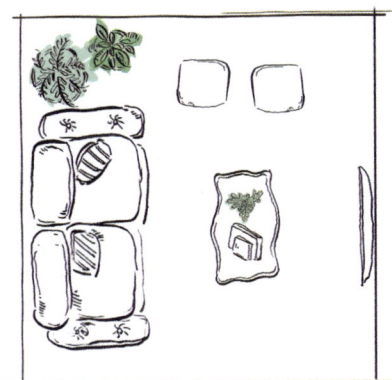

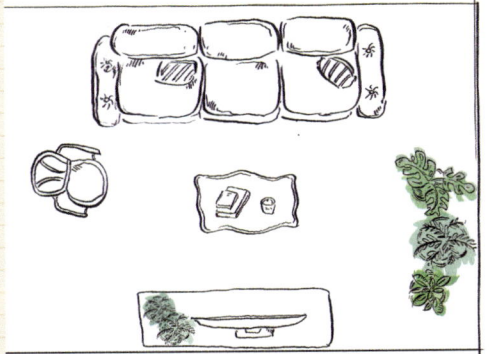

01 Place a pair of plants beside a sofa to soften the hard edges of a corner. Ensure that they have differing heights and foliage textures.

02 A line of three plants can help to break up large expanses of wall. Balance this with a smaller group of plants on a nearby cupboard or TV console.

LARGE LIVING ROOM

01 Even when your living room is full of furniture, plants can help to ground the space. They link together visually and unite the room elements.

02 Use plants to create harmony in your space. In this case, the two large plants in the corner balance the small plant on the coffee table in the middle of the room.

LARGE BEDROOM

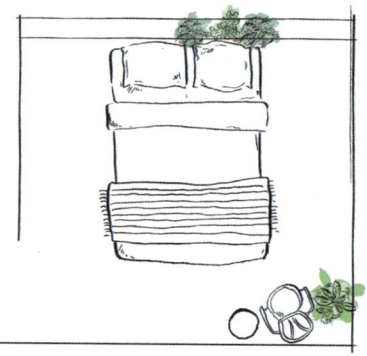

01 Wide bedheads are perfect for a line of small cascading and upright plants. Add another plant elsewhere in the room for balance.

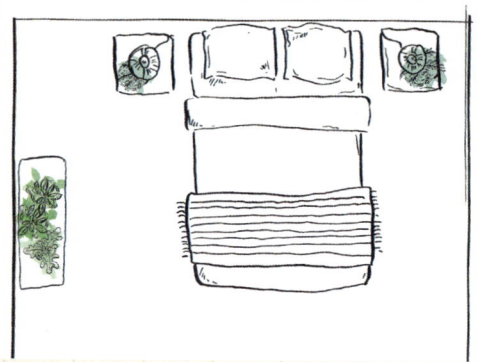

02 Capitalise on any unused flat surfaces in the bedroom. For example, incorporate a range of plants on your bedside tables and chest of drawers.

STANDARD BEDROOM

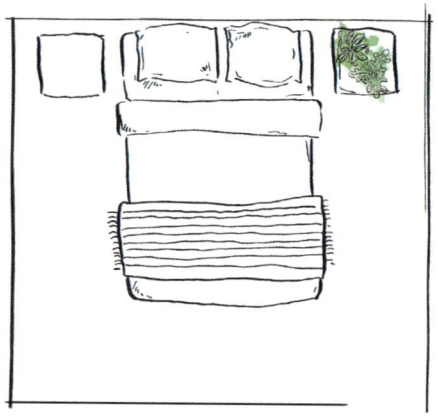

01 For an asymmetrical feel, place plants on just one bedside table. This adds an intriguing focal point to an otherwise uniform design.

SINGLE BEDROOM

01 To include greenery in a child's bedroom, take advantage of high spaces (such as the top of tall bookshelves).

02 Scale is vitally important for small bedrooms. You can still incorporate plants, but make sure that they're small or compact.

KITCHEN WITH ISLAND BENCH

01 A cluster of plants can help to soften stone or tile finishes on island benches. Just make sure that the plants are not in the way.

02 Deep corners on L-shaped kitchen benches can be filled with a range of greenery without taking up valuable working space.

GALLEY KITCHEN

 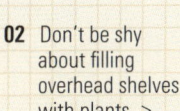 

01 Plants can instantly soften the abrupt ends of long benches and cabinets.

02 Don't be shy about filling overhead shelves with plants. >

KITCHENETTE

01 Overhead shelves create the perfect stage on which plants can dance.

02 A small addition of plants helps to mask the boring side of the pantry.

STUDY

01 Looking at a group of small plants — especially those with intricate or interesting foliage — will help to give your eyes a break from computer glare.

02 If you have enough space beside your desk, then you can bring the outside in — and freshen up the area — by incorporating a large plant with a canopy.

03 With working from home becoming a normal routine, our dining tables have become versatile. Adding a small cluster of plants will help to visually break up your table.

OUTDOORS

JULIETTE BALCONY

01 Place some hardy plants so you're not seeing bare concrete when looking out of the window.

02 On a narrow balcony, create an extended garden bed by using rectangular planters.

1-METRE (3-FOOT)-WIDE LONG BALCONY

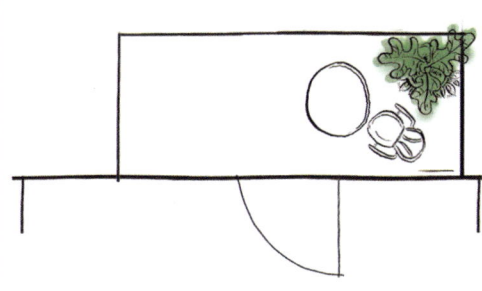

01 Frame the two outer corners of your balcony with clusters of plants. For visual interest, ensure that each group has a different number of plants.

02 Incorporate simple planting with outdoor furniture. A plant with height will help to create some shelter from the elements.

1-METRE (3-FOOT) SQUARE BALCONY

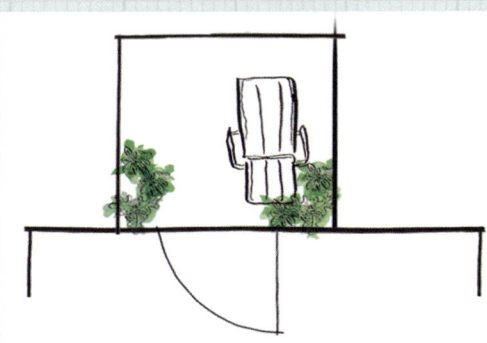

01 A cluster of planters of varying heights can create a beautiful moment to admire and will soften the boxy feel of the square space.

02 Group plants against the building to dissolve the hard boundary between the inside and the outside, and to encourage flow between the interior and exterior.

43

3-METRE (10-FOOT)-WIDE OR SQUARE BALCONY

01 Larger balconies benefit from a range of plant clusters. This helps to break up the expanse of timber or stone.

02 Style plants between furniture groupings to create moments of intimacy. Tall plants act like a screen, allowing people using the lounge chairs to have a little privacy.

ROOFTOP BALCONY

01 Exposure to the hot sun and drying winds is a common problem with rooftop balconies. It's best to group hardy arid and Mediterranean plants throughout your space, as they can cope with the harsh conditions.

1-METRE (3-FOOT)-WIDE SIDE GARDEN

01 The space between your house and the fence need not be desolate. Stepping stones provide a casual thoroughfare.

02 Even in narrow spaces, a series of climbing plants can make a big difference. Most grow up, not out, leaving plenty of room to move.

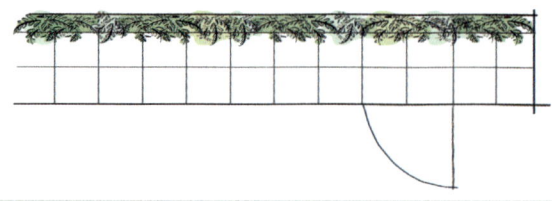

3-METRE (10-FOOT) X 4-METRE (13-FOOT) REAR GARDEN

01 Dense planting can create a sense of wonder and escape. Use a variety of foliage types to draw the eye across the garden.

02 A lush garden around your barbecue area can create the perfect outdoor room. Incorporate herbs that can be used in your cooking.

SUBURBAN FRONT YARD

01 A well-planted front yard is a welcoming gesture to visitors and will guide them to your front door. Choose plants with genuine street appeal.

02 You don't need to have a lawn. Feel free to fill your entire front yard with interesting plants – the native wildlife in your area will love it!

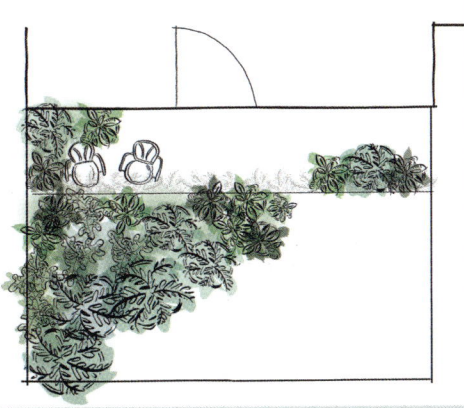

03 Use your verandah to incorporate more favourite plants. Potted plants help to smooth the transition from the outdoors to the indoors.

2-METRE (6-FOOT) X 4-METRE (13-FOOT) FRONT YARD

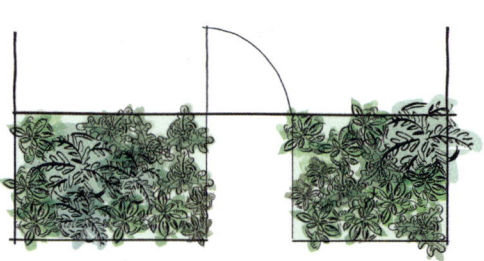

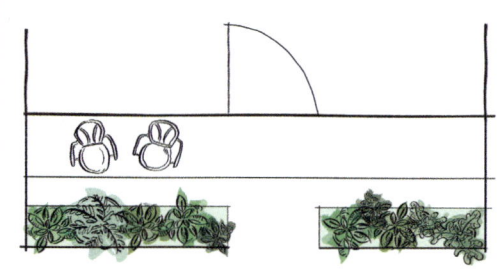

01 When densely planting a front yard, aim for interesting combinations. Otherwise, the garden becomes a mass of indistinct greenery.

02 If your property has a narrow front yard, then plant a strip of medium to tall plants to create privacy. It's much more appealing than building a wall or screen.

BARBECUE AREA

01 Use the space behind your barbecue to create a lush backdrop. Just make sure that the plants aren't too close to the cooking surface.

02 Soften the edges of a small barbecue with a group of three potted plants, and balance this with a trio of smaller plants on the table.

FRONT PORCH

01 Freshen up your seating arrangement with a touch of greenery. Having one plant on the left and two plants on the right offsets the symmetry of the furniture.

46

A SMALL GROUP OF PLANTS IN NEUTRAL-COLOURED PLANTERS CAN HELP TO BREAK UP A VISUAL CONFLICT BETWEEN COLOURS OR PATTERNS, MAKING THE SPACE EASIER ON THE EYE.

a guide to plants

Incorporating plants into our living spaces need not be daunting or confronting; the vast array of species and cultivars makes designing with plants a uniquely rich process. However, it's important to understand the materials you're working with. This allows you to design with more finesse and an ability to better foresee how your concepts will be realised. When it comes to plants, understanding their unique characteristics will help you to feel comfortable when designing with them.

UNDERSTANDING PLANTS AS THE KEY INGREDIENT

All artists have preferred media: for painters, it's paint plus canvas or paper. For gardeners, plants are the main medium. Planters, paving and architectural elements can all act as secondary media, but without plants, we don't have a garden.

Working out how to use a plant in a design can be a challenge. With an enormous range and variety of plants available, it can be overwhelming to curate the perfect combination.

When I started The Plant Society, and went from being an interior designer to a plant stylist, I was passionate about each plant adding a unique style to every project. Designers are trained in creating a vast array of styles and moods for their clients. Why shouldn't gardeners do the same?

When I look at a plant, I see it through the lens of an architect who loves gardening: to me, plants represent a certain aesthetic and can help to create an atmosphere. For example, the lush foliage of a dwarf umbrella tree (*Schefflera arboricola*) and fruit salad plants (*Monstera deliciosa*) gives a tropical feel (see opposite). So, you might ask, why should you choose certain plants, and what can the right plant add to your space?

51

Plant characteristics

When thinking about plant selection, consider the following key characteristics in relation to the architectural elements of your space (see also page 32).

GROWTH HABITS

A plant's typical growth habit is important when building a relationship between the scale of the plant and the space it will inhabit, particularly as it matures. Below are some common examples.

> **Climbing** – creeping fig (*Ficus pumila*), Virginia creeper (*Parthenocissus quinquefolia*), wax plants (*Hoya* spp.)
> **Clumping** – African arrowroot (*Canna indica*), bearded iris (*Iris × germanica*), peace lilies (*Spathiphyllum* spp.)
> **Dense** – *Casuarina glauca* 'Cousin It', pigfaces (*Carpobrotus* spp.)
> **Erect** – African milk tree (*Euphorbia trigona*), San Pedro cactus (*Trichocereus macrogonus* var. *pachanoi*)
> **Mat-forming** – creeping thyme (*Thymus serpyllum*), silver ponysfoot (*Dichondra argentea*)
> **Mound-forming** – Japanese lawn grass (*Zoysia japonica*), lavender cottons (*Santolina* spp.), mistletoe cacti (*Rhipsalis* spp.)
> **Open** – Madagascar palm (*Pachypodium lamerei*), Queensland bottle tree (*Brachychiton rupestris*)
> **Prostrate** – creeping rosemary (*Rosmarinus officinalis* Prostratus Group), orchid cacti (*Epiphyllum* spp.)
> **Spreading** – kidney weed (*Dichondra repens*), nasturtium (*Tropaeolum majus*), rock daisies (*Brachyscome* spp.)
> **Stemless** – aloe vera, poppies (*Papaver* spp.), primrose (*Primula vulgaris*)

FOLIAGE TEXTURE

> **Chunky foliage** – bush lily (*Clivia miniata*), fruit salad plant (*Monstera deliciosa*), tractor seat plant (*Cremanthodium reniforme*)
> **Fine foliage** – chain of hearts (*Ceropegia linearis* subsp. *woodii*), Delta maidenhair fern (*Adiantum raddianum*), Wollemi pine (*Wollemia nobilis*)
> **Glossy foliage** – gardenia, paperplant (*Fatsia japonica*), Zanzibar gem (*Zamioculcas zamiifolia*)
> **Matt foliage** – lamb's ear (*Stachys byzantina*), pig's ear (*Cotyledon orbiculata*), Spanish moss (*Tillandsia usneoides*)
> **Ruffled foliage** – *Cissus* 'Ellen Danica', hellebores (*Helleborus* spp.), kangaroo fern (*Microsorum pustulatum*)
> **Smooth foliage** – cast iron plant (*Aspidistra elatior*), Chinese money plant (*Pilea peperomioides*), rubber plant (*Ficus elastica*)

FOLIAGE TONE

> What colour is the leaf? Do you prefer a solid dark green or a song of colours? You may wish to create a minimalist look and restrict your tones to just shades of green, or you may want to select a few hues.

FOLIAGE PATTERN

> Does the leaf have a solid tone or is it variegated (differently coloured or patterned)? Variegation comes in a range of forms, from striped and marbled to spotted, which can help to add interest to your garden. Also note if the leaves have cut-outs that will help to create interesting shadows in your garden.

FLOWERING OR NOT?

> Does the plant produce noticeable blooms? Flowering plants provide any garden with a great sense of seasonality, and the flowers add unique colours and textures.

Breaking down your plant selection using these characteristics allows you to be strategic about why you are choosing each specific plant. The plants you select need to add to the bigger idea.

When I design, I separate each plant into groups based on these characteristics, then bring everything back together as a collection. This way, I can see individual details while also making sure that, in combination, the plants complement each other. Your plant palette works well as a combination when each plant adds a special quality to the overall collection. For example, your palette should have heroes – the feature plants – as well as a series of supporting plants that don't steal the show but act as important 'backup dancers' in your garden and establish a suitable backdrop.

Think about plants in terms of their design strengths. Each plant has a strong element that shapes the group in specific ways. Combining plants together is like weaving a tapestry. Each combination provides you with a different result. When it comes to gardens, why not celebrate them as artworks?

MY GO-TO PLANTS

As I've already mentioned, we are spoiled for choice when it comes to selecting plants for our gardens – the range available at garden centres and online is pretty mind-blowing. To help narrow it down, I've created a list of my favourite plants: staples that I use again and again, and that have been reliable choices for me in all types of gardens.

To achieve an organic flow in any green space, you need highs and lows (both visually and emotionally!). So, I often start with ground covers, then work my way through short and medium plants up to tall shrubs and trees. This ensures that I have a dynamic flow and shifts of height.

KEY TO PLANT CARE
☼ SUNLIGHT
◊ WATER
🌡 TEMPERATURE

Cascading plants and ground covers

Cascading plants and ground covers tie everything together. Consider these plants a buffer between other plant types, and between planters and the hard edges of furniture and surfaces. Cascading plants and ground covers are great for underplanting beneath larger plants; cascading plants can also be placed in pots indoors, on high shelves and on mantelpieces.

BLUE CHALK STICKS
(*Curio talinoides* var. *mandraliscae*)

☼ BRIGHT TO HARSH
◊ MODERATE
🌡 COOL TO WARM

CHAIN OF HEARTS
(*Ceropegia linearis* subsp. *woodii*)

☼ FILTERED TO HARSH
💧 MODERATE
🌡 COOL TO WARM

CISSUS **'ELLEN DANICA'**

☼ BRIGHT
💧 MODERATE
🌡 COOL TO WARM

COMMON BLUE VIOLET
(*Viola sororia*)

☼ FILTERED
💧 REGULAR
🌡 COOL TO WARM

CREEPING ROSEMARY
(*Rosmarinus officinalis* Prostratus Group)

☼ BRIGHT TO HARSH
💧 MODERATE
🌡 COOL TO WARM

CREEPING THYME
(*Thymus serpyllum*)

☼ BRIGHT TO HARSH
💧 MODERATE
🌡 COOL TO WARM

DICHONDRA ARGENTEA **'SILVER FALLS'**

☼ FILTERED TO BRIGHT
💧 REGULAR
🌡 COOL TO WARM

DONKEY'S TAIL
(*Sedum morganianum*)

☼ BRIGHT TO HARSH
💧 MODERATE
🌡 COOL TO WARM

HEARTLEAF PHILODENDRON
(*Philodendron hederaceum*)

☼ LOW TO BRIGHT
💧 MODERATE
🌡 WARM

KIDNEY WEED
(*Dichondra repens*)

☼ FILTERED TO BRIGHT
💧 REGULAR
🌡 COOL TO WARM

Cascading plants and ground covers

MISTLETOE CACTI
(*Rhipsalis* spp.) –
fine leaf and flat leaf

☼ BRIGHT
◊ SPARINGLY
🌡 COOL TO WARM

MONKEY TAIL CACTUS
(*Cleistocactus winteri* subsp.
colademono)

☼ HARSH
◊ RARELY
🌡 COOL TO WARM

NATIVE VIOLET
(*Viola hederacea*)

☼ FILTERED
◊ REGULAR
🌡 COOL TO WARM

ORCHID CACTI
(*Epiphyllum* spp.)

☼ BRIGHT
◊ SPARINGLY
🌡 COOL TO WARM

PIGFACES
(*Carpobrotus* spp.)

☼ BRIGHT TO HARSH
◊ SPARINGLY
🌡 COOL TO WARM

SEDUM 'BLUE FEATHER'

☼ BRIGHT TO HARSH
◊ MODERATE
🌡 COOL TO WARM

SIKKIM CREEPER
(*Parthenocissus sikkimensis*)

☼ FILTERED TO BRIGHT
◊ MODERATE
🌡 COOL TO WARM

STRING OF BANANAS
(*Curio radicans*)

☼ BRIGHT TO HARSH
◊ MODERATE
🌡 COOL TO WARM

WAX PLANTS
(*Hoya* spp.)

☼ FILTERED TO BRIGHT
◊ MODERATE
🌡 WARM

Climbing plants

If your space is small, why not go up? Climbing plants are ideal for places where your ground area is limited, and they can also be used to soften the view of hard surfaces, such as walls or fences. There is a great selection of either suckering or vining climbing plants that will happily grow upwards.

BOSTON IVY
(*Parthenocissus tricuspidata*)

☼ BRIGHT TO HARSH
◊ MODERATE
🌡 COOL TO WARM

CHESTNUT VINE
(*Tetrastigma voinierianum*)

☼ FILTERED TO BRIGHT
◊ MODERATE
🌡 COOL TO WARM

CHINESE WISTERIA
(*Wisteria sinensis*)

☼ BRIGHT
◊ MODERATE
🌡 COOL TO WARM

CLIMBING HYDRANGEA
(*Hydrangea petiolaris*)

☼ FILTERED
◊ MODERATE
🌡 COOL TO WARM

CLIMBING ROSES

☼ BRIGHT
◊ MODERATE
🌡 COOL TO WARM

CREEPING FIG
(*Ficus pumila*)

☼ FILTERED TO HARSH
◊ REGULAR
🌡 COOL TO WARM

DEVIL'S IVY
(*Epipremnum aureum*)

☼ LOW TO BRIGHT
◊ MODERATE
🌡 WARM

Climbing plants

DRAGON-TAIL PLANT
(*Epipremnum pinnatum*)

☼ FILTERED TO BRIGHT
◊ MODERATE
🌡 WARM

GRAPE VINE
(*Vitis vinifera*)

☼ BRIGHT
◊ MODERATE
🌡 COOL TO WARM

KIWIFRUIT
(*Actinidia chinensis* var. *deliciosa*)

☼ BRIGHT
◊ MODERATE
🌡 COOL TO WARM

VIRGINIA CREEPER
(*Parthenocissus quinquefolia*)

☼ BRIGHT TO HARSH
◊ MODERATE
🌡 COOL TO WARM

WAX PLANTS
(*Hoya* spp.)

☼ FILTERED TO BRIGHT
◊ MODERATE
🌡 WARM

Short plants

Using short plants in your garden helps to establish a strong ground layer. They can also be used to soften edges and lines. Small plants create intimate moments in a garden – they require us to pause and look closely at their foliage and flowers.

BABY RUBBER PLANT
(*Peperomia obtusifolia*)

☼ FILTERED TO BRIGHT
◊ MODERATE
🌡 WARM

Short plants

BEE BALMS
(*Monarda* spp.)

☼ BRIGHT TO HARSH
◊ MODERATE
🌡 COOL TO WARM

BLUE STAR FERN
(*Phlebodium aureum*)

☼ FILTERED TO BRIGHT
◊ MODERATE
🌡 COOL TO WARM

BOSTON FERN
(*Nephrolepis exaltata*)

☼ FILTERED TO BRIGHT
◊ REGULAR
🌡 COOL TO WARM

CARDBOARD PALM
(*Zamia furfuracea*)

☼ BRIGHT TO HARSH
◊ SPARINGLY
🌡 WARM

CHINESE MONEY PLANT
(*Pilea peperomioides*)

☼ FILTERED TO BRIGHT
◊ MODERATE
🌡 COOL TO WARM

DAFFODILS
(*Narcissus* spp.)

☼ FILTERED TO HARSH
◊ MODERATE
🌡 COOL

DAHLIAS
(*Dahlia pinnata*)

☼ BRIGHT
◊ REGULAR
🌡 WARM

FAN ALOE
(*Kumara plicatilis*)

☼ BRIGHT TO HARSH
◊ SPARINGLY
🌡 WARM

GIANT SWORD FERN
(*Nephrolepis biserrata*)

☼ FILTERED TO BRIGHT
◊ REGULAR
🌡 COOL TO WARM

Short plants

LAVENDER COTTON
(*Santolina chamaecyparissus*)

☼ BRIGHT TO HARSH
○ SPARINGLY
🌡 COOL TO WARM

MICHAELMAS DAISIES
(*Aster* spp.)

☼ BRIGHT TO HARSH
○ MODERATE
🌡 COOL TO WARM

PIG'S EARS
(*Cotyledon* spp.)

☼ BRIGHT TO HARSH
○ SPARINGLY
🌡 COOL TO WARM

SEDGES
(*Carex* spp.) –
green and bronze

☼ BRIGHT TO HARSH
○ SPARINGLY
🌡 COOL TO WARM

SNAKE PLANTS
(*Dracaena* spp.)

☼ FILTERED TO BRIGHT
○ SPARINGLY
🌡 COOL TO WARM

SPURGES
(*Euphorbia* spp.)

☼ BRIGHT TO HARSH
○ SPARINGLY
🌡 COOL TO WARM

STAGHORN FERNS
(*Platycerium* spp.)

☼ FILTERED
○ REGULAR
🌡 COOL TO WARM

STONECROPS
(*Sedum* spp.)

☼ BRIGHT TO HARSH
○ SPARINGLY
🌡 COOL TO WARM

TRACTOR SEAT PLANT
(*Cremanthodium reniforme*)

☼ FILTERED TO BRIGHT
○ SPARINGLY
🌡 COOL TO WARM

WINTER ROSE
(*Helleborus orientalis*)

☼ FILTERED TO BRIGHT
◊ MODERATE
🌡 COOL

YARROWS
(*Achillea* spp.)

☼ BRIGHT TO HARSH
◊ MODERATE
🌡 COOL TO WARM

ZANZIBAR GEM
(*Zamioculcas zamiifolia*)

☼ LOW TO BRIGHT
◊ SPARINGLY
🌡 WARM

Medium plants

Medium plants create a great transition between short and tall plants, guiding our gaze upwards from the ground. Although they help to fill out the middle plane of your garden, they don't necessarily need to be overpowering. Outdoors, delicate grasses – such as reed grasses (*Calamagrostis* spp.) – create a whimsical atmosphere. Indoors, plants such as dwarf umbrella tree (*Schefflera arboricola*) add a gentle touch of green to your interior.

AFRICAN MILK TREE
(*Euphorbia trigona*)

☼ BRIGHT TO HARSH
◊ RARELY
🌡 WARM

BUTTERFLY BUSHES
(*Buddleja* spp.)

☼ BRIGHT TO HARSH
◊ MODERATE
🌡 COOL TO WARM

CHINESE SILVER GRASS
(*Miscanthus sinensis*)

☼ BRIGHT TO HARSH
◊ MODERATE
🌡 COOL TO WARM

DWARF UMBRELLA TREE
(*Schefflera arboricola*)

☼ FILTERED TO BRIGHT
◊ MODERATE
🌡 WARM

Medium plants

FENNEL
(*Foeniculum vulgare*) –
green and bronze

☼ BRIGHT TO HARSH
◊ MODERATE
🌡 COOL TO WARM

FRUIT SALAD PLANT
(*Monstera deliciosa*)

☼ FILTERED TO BRIGHT
◊ MODERATE
🌡 COOL TO WARM

PANIC GRASSES
(*Panicum* spp.)

☼ BRIGHT TO HARSH
◊ MODERATE
🌡 COOL TO WARM

PEACE LILIES
(*Spathiphyllum* spp.)

☼ FILTERED TO BRIGHT
◊ MODERATE TO REGULAR
🌡 WARM

REED GRASSES
(*Calamagrostis* spp.)

☼ BRIGHT TO HARSH
◊ MODERATE
🌡 COOL TO WARM

SAGES
(*Salvia* spp.)

☼ FILTERED TO BRIGHT
◊ REGULAR
🌡 COOL TO WARM

Tall plants

Tall plants add a sense of grandeur to your space. They are
also great for breaking up large indoor and outdoor areas,
so you're left with smaller, more intimate zones.

CERCIS CANADENSIS
'FOREST PANSY'

☼ FILTERED TO BRIGHT
◊ MODERATE
🌡 COOL TO WARM

Tall plants

CITRUSES

☼ BRIGHT TO HARSH
◊ MODERATE
🌡 COOL TO WARM

FIDDLE LEAF FIG
(*Ficus lyrata*)

☼ FILTERED TO BRIGHT
◊ MODERATE
🌡 WARM

**GIANT WHITE BIRD
OF PARADISE**
(*Strelitzia nicolai*)

☼ BRIGHT
◊ MODERATE
🌡 COOL TO WARM

MAIDENHAIR TREE
(*Ginkgo biloba*)

☼ BRIGHT TO HARSH
◊ MODERATE
🌡 COOL TO WARM

MOUNTAIN CABBAGE TREE
(*Cussonia paniculata*)

☼ BRIGHT TO HARSH
◊ MODERATE
🌡 COOL TO WARM

RUBBER PLANT
(*Ficus elastica*)

☼ FILTERED TO BRIGHT
◊ MODERATE
🌡 COOL TO WARM

SAN PEDRO CACTUS
(*Trichocereus macrogonus*
var. *pachanoi*)

☼ BRIGHT TO HARSH
◊ RARELY
🌡 WARM

TREE ALOE
(*Aloidendron barberae*)

☼ BRIGHT TO HARSH
◊ SPARINGLY
🌡 COOL TO WARM

UMBRELLA TREE FIG
(*Ficus umbellata*)

☼ BRIGHT TO HARSH
◊ MODERATE
🌡 WARM

63

Trees

Trees make a statement – they command a space and add a clear hierarchy to your garden. Even in a small area, you can include a potted tree to create a strong presence.

CREPE MYRTLES
(*Lagerstroemia* spp.)

☀ BRIGHT TO HARSH
◊ MODERATE
🌡 COOL TO WARM

EVERFRESH TREE
(*Albizia splendens*)

☀ BRIGHT TO HARSH
◊ REGULAR
🌡 COOL TO WARM

KAURI PINE
(*Agathis robusta*)

☀ BRIGHT TO HARSH
◊ MODERATE
🌡 COOL TO WARM

MAPLES
(*Acer* spp.)

☀ BRIGHT TO HARSH
◊ MODERATE
🌡 COOL TO WARM

OLIVE TREE
(*Olea europaea*)

☀ BRIGHT TO HARSH
◊ MODERATE
🌡 COOL TO WARM

QUEENSLAND BOTTLE TREE
(*Brachychiton rupestris*)

☀ BRIGHT TO HARSH
◊ MODERATE
🌡 COOL TO WARM

***SCHEFFLERA ACTINOPHYLLA*
'AMATE'**

☀ FILTERED TO BRIGHT
◊ MODERATE
🌡 WARM

SILK FLOSS TREE
(*Ceiba speciosa*)

☀ BRIGHT TO HARSH
◊ MODERATE
🌡 COOL TO WARM

WEEPING FIG
(*Ficus benjamina*)

☀ BRIGHT TO HARSH
◊ REGULAR
🌡 COOL TO WARM

THE ART OF GROUPING PLANTS

Working out what plants go with what can be quite challenging. However, being a gardener is about trial and error; there is joy in experimentation. Through practice, I've discovered that certain plants work well together. But I'm still constantly trying to push my skill set. Some combinations work, some don't – it's a continuous learning curve.

Start by breaking down the individual qualities of each plant in your Kit of Parts (such as size, colour, foliage type and growth habit; see pages 52–3), before looking at its relationship to other plant types. How does it connect to or contrast with other plants? For example, does its variegated foliage stand out in a sea of green? Will it form a 'red carpet' for other stars in your garden? Can its height or shape be used to add drama to an otherwise calm vignette? Once you understand your plants a little better, it's time to group them to create a coherent atmosphere and sense of place.

Begin by selecting the plant that you want to lead your design. This may be a focus tree, if you're using one, or it could be a plant that has become a favourite because of the foliage it bears or the form it has. Your lead plant will be special to you and is usually the plant that draws you in the most.

Then layer in the other plants (for detailed instructions on how to do this, see page 98), aiming to find ways for each to complement the others.

Think of it like cooking a meal: good recipes require a few ingredients to create balance, and sometimes a bold element may need to be softened with another ingredient. Consider how all of your plant choices can create balance in your space. My tip is to gradually build up your selection, rather than going in hard with everything at once.

Successfully grouping plants is a skill that can be developed with practice. To give you a head start, opposite are some examples of my favourite combinations.

CHUNKY FOLIAGE IS BOLD AND EYE-CATCHING, WHILE FINE FOLIAGE CARESSES AND FILLS THE GAPS. PAIRING THE TWO FOLIAGE TYPES CREATES AESTHETIC MOVEMENT IN THE GARDEN.

FICUS ELASTICA 'BURGUNDY'

plus **kidney weed**
   (*Dichondra repens*)

The bold, dark, architectural leaves of *Ficus elastica* 'Burgundy' – one of my favourites – are softened by the small, delicate leaves of kidney weed. When the foliage of kidney weed cascades over the planter, this further helps to balance the organic and the manufactured.

kidney weed

Ficus elastica 'Burgundy'

KAURI PINE
(*Agathis robusta*)

plus **Chinese money plant**
   (*Pilea peperomioides*)
plus ***Cissus* 'Ellen Danica'**

Tall-growing kauri pine brings height to the trio. It also bears small, oval-shaped leaves that complement those of Chinese money plant and *Cissus* 'Ellen Danica'. The mix of leaf shapes creates an interesting layering of elegant foliage.

Chinese money plant

kauri pine

Cissus 'Ellen Danica'

CITRUS

plus **creeping thyme**
   (*Thymus serpyllum*)

Here's a simple yet classic pairing. Both have rich green leaves, with a transition from the glossy foliage of citrus to the delicate carpet of fine foliage offered by creeping thyme.

citrus

creeping thyme

TRACTOR SEAT PLANT
(*Cremanthodium reniforme*)

plus **green sedge** (*Carex viridula*)

plus **Chinese silver grass**
(*Miscanthus sinensis*)

The bold and glossy foliage of tractor seat plant grounds the trio, while the soft and airy textures of green sedge and Chinese silver grass add a whimsical touch.

tractor seat plant

green sedge

Chinese silver grass

SAGE
(*Salvia officinalis*)

plus **native violet**
(*Viola hederacea*)

plus **reed grass**
(*Calamagrostis* sp.)

plus **lamb's ear**
(*Stachys byzantina*)

This mix of leaf textures transitions from the rich green tones of sage and native violet to the dusty silvers found in reed grass and lamb's ear. Incorporating a range of subtle shifts in foliage tone can create a sophisticated sense of movement.

sage

native violet

reed grass

lamb's ear

YARROW

(*Achillea* sp.)

plus **Carex comans 'Bronze'**

plus **Haloragis erecta 'Wellington Bronze'**

plus **tulip bulbs**

This play of delicate leaf forms bears burnt and dusty tones. It's a delicate grouping that is seasonally punctuated with the rise of tulip foliage and flowers.

Carex comans 'Bronze'

yarrow

tulip bulbs

Haloragis erecta 'Wellington Bronze'

CABBAGE TREE

(*Cussonia* sp.)

plus **Dichondra argentea 'Silver Falls'**

plus **Ficus elastica 'Ruby'**

plus **maidenhair tree**
(*Ginkgo biloba*)

Here is a series of sculptural specimens. Each has unique leaf colours, creating a contrast between the silver foliage of cabbage tree and *Dichondra argentea* 'Silver Falls', and the maroon foliage of *Ficus elastica* 'Ruby'. With its light green, fan-shaped leaves, maidenhair tree is an iconic backdrop to the grouping.

cabbage tree

Dichondra argentea 'Silver Falls'

Ficus elastica 'Ruby'

maidenhair tree

CHAPTER THREE

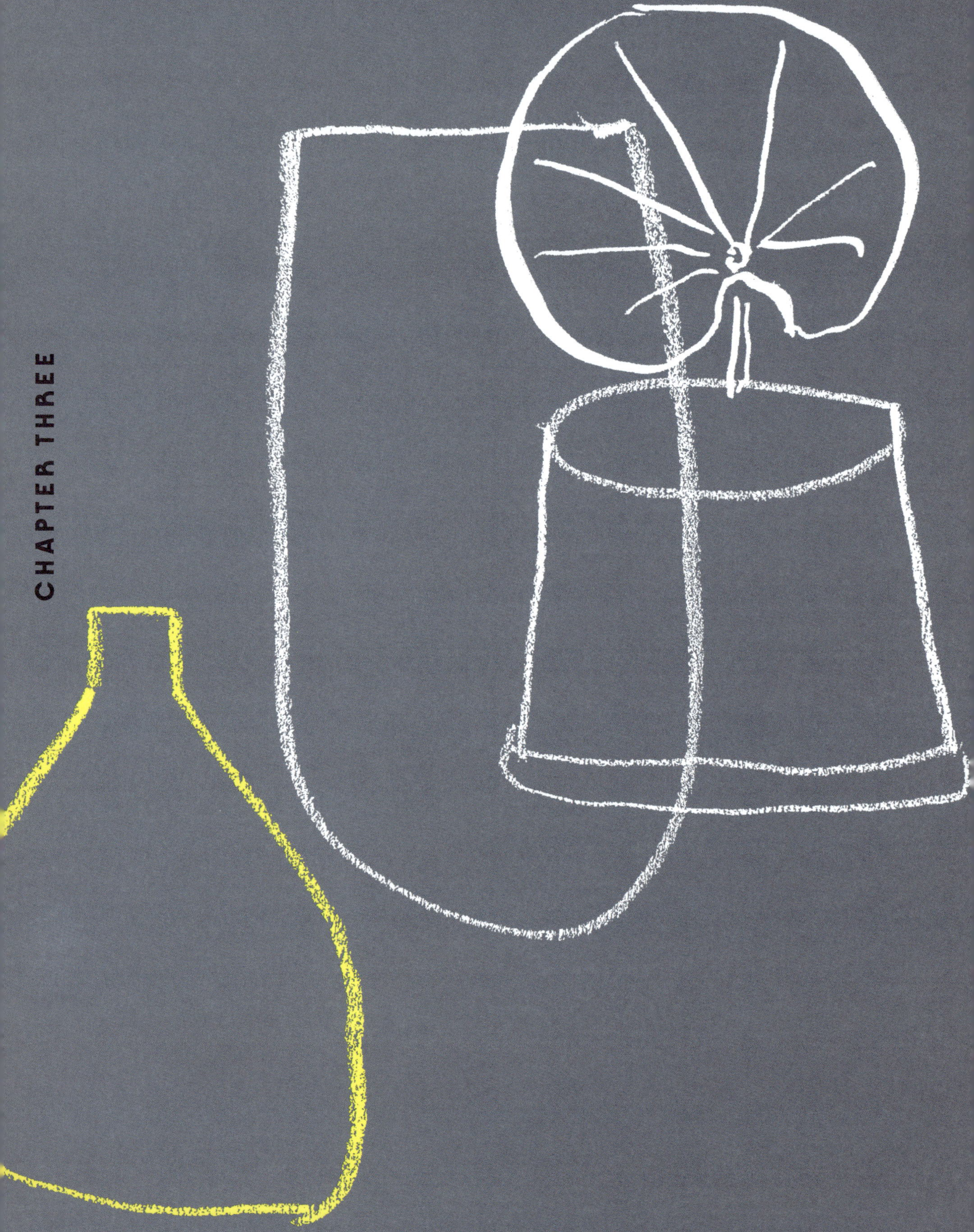

designing with plants

When designing with plants, I like to consider each individual element, and look at how it relates to core design fundamentals, which are the building blocks of my work. This approach helps to create a coherent look – an atmosphere – through the space. Atmosphere is what makes a green space unique – I want to be able to see the gardener in the space they've created.

DESIGN
FUNDAMENTALS

On the following pages I share my tips on creating a design-driven green space for your home. We can't all be expert designers; sometimes it's just about having the right ingredients and knowing what to do with them. I want to show you how to express your unique style through these key components of design:

> **textures and shapes**
> **colour**
> **scale**
> **vignettes**

> **styling your space**
> **plant placement**
> **connecting all**
  the elements.

Textures and shapes

When you think about how your plants will interact with the space you're designing, ensure that you consider the part that the texture and shape of the planters will play.

Planter textures and materials

Texture is my favourite design tool. When you work on an interior or garden, consider how natural and manufactured textures can either complement each other or clash intentionally.

Planters come in a wide variety of materials, and each offers a unique texture for your space. Sometimes it's the imperfections in a design that I love, and the introduction of patinaed or weathered planters adds another element of personality. Here are some suggestions for plants that suit particular planters.

CERAMIC

Smooth gloss

> To balance glossy materials, pair them with matt foliage (the hairy foliage of begonias makes them well suited for this combination). The matt foliage helps to tone down the glossy appearance of the planter, while light bounces off the planter's surface to brighten the space.

Smooth matt

> Waxy foliage, from glossy succulents and wax plants (*Hoya* spp.), works particularly well with smooth matt planters, as the matt material creates a subtle backdrop to the foliage. The material is also suited to the grey foliage found on the olive tree (*Olea europaea*) and hairy plants such as old man cactus (*Cephalocereus senilis*).

Etched

> Plants with upright foliage – such as snake plants (*Dracaena* spp.) and twin-flowered agave (*Agave geminiflora*) – work well with detailed planters. Their upright nature allows the planter detail to remain visible.

Fine flute

> I love using textures that catch light and shadow. Fine fluting provides a delicate canvas for light to bounce off. Chunky foliage – such as that of felt bush (*Kalanchoe beharensis*) or fruit salad plant (*Monstera deliciosa*) – pairs well with such fine textures because the foliage doesn't compete with the planter.

Wide flute

> With a larger scale of texture, you can go even simpler with the foliage you choose. Plants such as cacti have uncluttered forms that stand strong in wide-fluted planters. Alternatively, if you want your garden to be texture-heavy, you can pair grasses and other intricately textured foliage types with wide-fluted planters.

Patina

> I'm a huge fan of patina and worn finishes. They tell a story of the past and help to bring nostalgia to a project. With patina finishes, the world is your oyster when it comes to pairing foliage tones. Using gnarly plants – such as aged olive trees (*Olea europaea*) – and more free-flowing plant forms adds even more atmosphere to your planting.

TERRACOTTA
Various

> I reserve terracotta for Mediterranean and arid plants, such as pomegranate (*Punica granatum*) and fan aloe (*Kumara plicatilis*). Not only are the plant types suited to the dry environment created by the porous nature of terracotta, but they also tell a story of sun-kissed climates.

STYLING TIP
WHEN USING A FEW
DIFFERENTLY SHAPED
PLANTERS, KEEP THE
OVERALL MOOD CALM
BY HAVING SIMILAR
COLOURS OR TONES
ACROSS ALL OF
THE PLANTERS.

FABRIC

Canvas

> With the simple form of a canvas planter, try pairing it with bold plants such as fruit salad plant (*Monstera deliciosa*) or prickly pears (*Opuntia* spp.). Both the planter and the plant will have a strong silhouette that makes for a bold look.

Felted

> The softness of felted planters lends itself to delicate plants, such as ferns or chain of hearts (*Ceropegia linearis* subsp. *woodii*). The pairing of the two creates a gentle mood.

STONE

Marble

> Interiors the world over feature marble, so why not incorporate marble planters for a touch of elegance? Paired with finer leaf forms – such as the foliage found on *Cissus* 'Ellen Danica' – this combination conveys an understated level of class.

Rustic

> Heavily fired planters with rough, charred finishes are a moody addition to your garden. As they are typically very textured, pair them with bold foliage – such as that of tree aloe (*Aloidendron barberae*) – to ensure that the texture remains visible.

WOODY

Woven

> Channelling a more relaxed look, woven baskets – whether vintage or new – can help to calm or bring charm to an interior or outdoor space. Indoors, the large and leafy foliage of fiddle leaf fig (*Ficus lyrata*) creates a great balance between intricate and bold. Outdoors, baskets of bulbs work well together to create seasonal highlights.

Smooth timber

> Refined timber planters call for heavily textured foliage. Dwarf umbrella tree (*Schefflera arboricola*) is a great pairing. Its delicate leaves cast intricate shadows against the smooth timber planter.

Cork

> A material with such a detailed texture calls for either a simple approach, by pairing it with Chinese money plant (*Pilea peperomioides*) or a similar species, or a heavily textured approach, by using spiderworts (*Tradescantia* spp.), for example.

Planter shapes and forms

With endless planter options at your fingertips, you're spoiled for choice – and it's quite easy to become overwhelmed. When considering shapes and forms, there are certain planters that work better in specific scenarios and environments. You can also source found objects – such as baskets, vases or even crockpots – and repurpose them as planters. There are no rules when it comes to gardening! Below is a guide to using different planter shapes and forms.

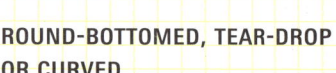

ROUND-BOTTOMED, TEAR-DROP OR CURVED

> These are suitable for a wide range of applications, but should be avoided on balconies (as balconies tend to be windy places, the curved planter bottoms can cause the plants to topple over). If you have children or pets, it's important to consider the stability of your planters. Curved and round-bottomed planters can easily tip over when pulled or knocked. I recommend doing a shake test to see how easily the planter of choice can be tipped over.

CYLINDRICAL

> Round, upright planters with a flat base are great for a variety of situations. They create simple silhouettes and are much sturdier than round-bottomed, tear-drop or curved planters.

TAPERED

> These provide subtle detail and a sophisticated touch to your garden. They can add some height to elevate your plants.

NARROW

> If you have a tight space, then a narrow planter is ideal.

TALL

> These planters can help to increase the height of your plants and provide a canvas for a cluster of smaller planters in front.

SHORT

> If you want to create a low landscape or maintain a view, then use short and shallow planters. They allow you to add greenery while preserving a valuable line of sight.

SAUCER OR NO SAUCER?

Saucers protect floors and other surfaces in your home from water damage, but they're not always aesthetically pleasing. Here are three approaches to saucer use:

1. Use a saucer with the same tone and texture as the planter so it blends in. Place felt or rubber dots on the bottom to allow airflow and minimise condensation.
2. Adding a saucer can ruin the silhouette of the planter. In this case, seal the drainage holes with a waterproofing agent and use the planter as a decorative cover. Simply place a drip tray into the bottom of the planter, keep your plant in a grower's planter (make sure it's slightly smaller than the decorative planter), and insert the grower's planter into the decorative planter.
3. I'm not worried about floor damage outdoors, so I generally don't use saucers. To promote airflow under the planter, I use pot feet.

Colour

In addition to using coloured flowers and foliage, there are many ways you can introduce different hues into your garden – think planters and other materials or objects, such as paving, painted surfaces and furniture.

Feature colour

Using colour in gardens is very personal. When it comes to choosing a colour for planters, I prefer earthy hues, as they are less trend-driven. They also allow foliage to stand out, rather than compete with the planter for attention. Grittier tones create more of a connection to nature.

When selecting the colours for your project, start by choosing the primary feature colour for the materials in your space. Then pair this with the foliage you want to use in your design. Here are some examples.

#1 RUSSET
RUBBER PLANT
(*FICUS ELASTICA*)

#2 NUTMEG
MONKEY TAIL CACTUS
(*CLEISTOCACTUS WINTERI
SUBSP. COLADEMONO*)

#3 HUNTER GREEN
ZIGZAG CACTUS
(*DISOCACTUS
ANGULIGER*)

#4 CHARCOAL
TRACTOR SEAT PLANT
(*CREMANTHODIUM
RENIFORME*)

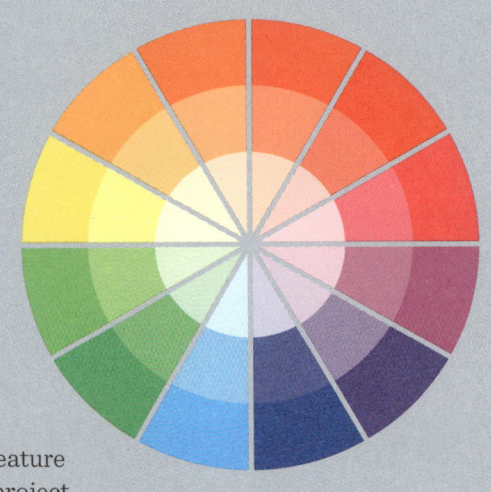

Complementary colours

Some projects require more than one colour. After selecting your feature colour, you can choose complementary colours that also suit your project. Lay these colours side by side in situ, as shades change depending on the lighting in the space. (When looking at paint finishes, it helps to have physical colour swatches; you can get these from your local hardware or paint store.)

Make sure to consider all of the colours in your space, including those of furniture, walls and homewares. Collect samples of any materials that will cohabit in the space – such as metals, timbers and soft furnishings – and lay them out alongside your chosen colours. There may be existing colours in your space that are at odds with your intended palette, in which case you might need to revisit and adjust your colour scheme.

IF YOU'RE STRUGGLING TO SOURCE PLANTERS THAT SUIT YOUR COLOUR PALETTE, THEN CONSIDER FINDING A PLANTER THAT IS THE RIGHT FORM AND PAINTING IT TO FIT THE AESTHETIC.

SAND STORM

RUSSET

BROWN CLAY

My favourite colour combinations

| BURNT ORANGE | BURGUNDY | RATTAN | CLAY/SANDY | ORANGE NEW ZEALAND SEDGE (*CAREX TESTACEA*) | *ROSA* 'FRUITY PARFUMA' | WINTER ROSE (*HELLEBORUS ORIENTALIS*) |

Earthy This palette exudes warmth and gives the space an inviting ambience.

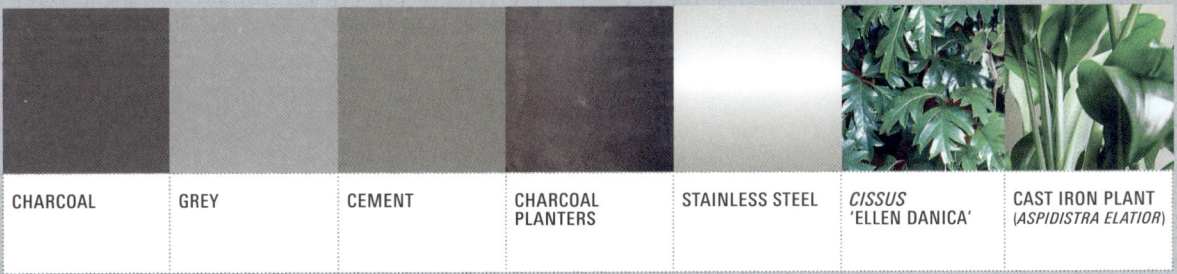

| CHARCOAL | GREY | CEMENT | CHARCOAL PLANTERS | STAINLESS STEEL | *CISSUS* 'ELLEN DANICA' | CAST IRON PLANT (*ASPIDISTRA ELATIOR*) |

Silvery Cool and sophisticated, this neutral palette suits a modern home.

| FRENCH BLUE | MARBLE | DARK TIMBER | STAINLESS STEEL | BROWN CLAY | BLUE STAR FERN (*PHLEBODIUM AUREUM*) | CYLINDRICAL SNAKE PLANT (*DRACAENA ANGOLENSIS*) |

Organic Inspired by natural materials, this palette boasts old-world charm.

| RUSSET | SALMON | TERRACOTTA | FOXTAIL AGAVE (*AGAVE ATTENUATA*) | COMMON BUGLE (*AJUGA REPTANS*) | SAGE (*SALVIA SP.*) |

Tuscan The rich and rustic colours in this palette will transport you straight to Italy.

Colour combinations
CONTINUED

| LIGHT BROWN | RENDER | OAK TIMBER | SANDSTONE | CUMQUAT TREE (*CITRUS JAPONICA*) | CABBAGE TREE (*CUSSONIA SP.*) | CREEPING THYME (*THYMUS SERPYLLUM*) |

Coastal You can almost smell the sea breeze with this beachy palette.

| MUSTARD | BROWN | OFF-WHITE PLANTERS | STONE | *BUDDLEJA DAVIDII* 'WHITE PROFUSION' | SALTBUSH (*ATRIPLEX HALIMUS*) | LAVENDER COTTON (*SANTOLINA* SP.) |

Sunny Yellows shine brightly among the silvery tones in this pretty palette.

| SAGE GREEN | AGED TIMBER | HEARTLEAF PHILODENDRON (*PHILODENDRON HEDERACEUM*) | BABY RUBBER PLANT (*PEPEROMIA OBTUSIFOLIA*) | ZANZIBAR GEM (*ZAMIOCULCAS ZAMIIFOLIA*) |

Leafy Green on green with a hint of wood makes for a superb sylvan palette.

| WHITE | WHITE LINEN | AGED TERRACOTTA | POWDER-COATED METAL | OLIVE TREE (*OLEA EUROPAEA*) | CREEPING ROSEMARY (*ROSMARINUS OFFICINALIS* PROSTRATUS GROUP) | FAN ALOE (*KUMARA PLICATILIS*) |

Mediterranean This pleasing palette adds balmy holiday vibes to any space.

Scale

Scale is an important element to consider when designing because our eyes find a natural pleasure in balanced creations and displays. When designing with plants, it's essential to consider the scale of a plant (or your garden) in relation to the space you're designing. There are two main principles that will assist you in setting up the scale of your designs.

Principle 1

Use a hierarchy of scale. Our eyes adjust easily to points, which means that when we view a collection of objects, they are easier on the eye when they stack up to one point.

Principle 2

Break up the space into thirds. When we look at landscape paintings or photographs, the images feel more natural when they are cropped to accommodate the idea of thirds.

01 Styling plants to a point creates a sense of order. The top of the tallest plant is the focal point and provides a clear starting place from which to view the rest of the grouping. Our eyes are drawn downwards towards the complexity of the cluster.

02 Our eyes find comfort in the horizon line and when spaces feel organic. Breaking each space into imaginary thirds helps to re-create the horizon line.

STYLING TIP
WHEN CONSIDERING
SCALE IN YOUR DESIGN,
AVOID FOCUSING ON THE
DETAILS OF THE PLANTS
AND PLANTERS YOU
PROPOSE TO USE;
INSTEAD, THINK ABOUT
THEIR SILHOUETTES.

Vignettes

Beginning your garden project can seem like a huge mission. So, try to break down the project into manageable segments (known as vignettes). Tackling smaller components will allow you to style without being overwhelmed. It will also make it easier to achieve a balanced hierarchy in your space and, in turn, give your space the emotional highs and lows it requires.

However, always remember that working on individual spaces shouldn't come at the expense of the entire aesthetic. Continue to use your Kit of Parts (see page 30) as a guide and benchmark for your overall atmosphere and to ensure that you maintain a unified style. Consider it a design compass that will prevent you from straying from your established look and feel.

How to begin

Think about the main viewpoints and work from there. For instance, the view out to your balcony from your living room may be your most important line of sight, so why not begin by curating a garden that can be enjoyed from the place you spend the most time? Start designing from this angle and let the creativity flow from there.

Once you have curated a vignette in your space, use this to inspire your next vignette. Don't be afraid to continue using the same materials and plants in other areas. Repetition is a great way to build continuity. To create a slight uniqueness in each area, you can incorporate the odd individual piece. I like using handmade or vintage items so that each space has a special 'moment'.

Styling your space

Be creative when seeking out materials for your design so you can make it unique. I like to source from a range of stores to find the right plants and planters for my projects. It's important to branch out, rather than buying everything from the main homeware stores and garden centres – otherwise your design may feel like a cookie-cutter project. Below are some key places from where you can source interesting materials.

CHAIN STORES

> These are a go-to because they have a broad range of planters and objects. I like to source more classic items from these stores. If you steer clear of trending items, then your designs will age well. Source materials that form a patina over time – such as terracotta – to ensure that your project doesn't feel 'plastic'.

LOCAL MARKETS

> A great way to add character to your design is to incorporate some locally produced pieces. Although more costly, these handmade items introduce a point of difference. Make sure to place these special pieces where they are most visible, such as on tabletops and windowsills.

HOMEWARE STORES

> Niche homeware stores located off the high streets or online are excellent resources for finding unique pieces. They often stock brands that make bespoke products in limited runs.

VINTAGE STORES

> These allow you to step back in time. With the right eye, you'll find pieces that relate to your concept.

ON THE STREET

> I keep finding things on the side of the street, thrown out during a move or no longer wanted by their owner. Recycling or re-using found pieces can add a whole new layer of interest to your garden.

SOCIAL MEDIA

> Don't hesitate to look on social media for second-hand goods. These can range from vintage to simply no longer wanted (but perfect for your garden).

PROFESSIONAL ARTISANS

> Why not have a ceramicist or woodworker create a one-off piece for your garden? It will become a talking point!

STYLING TIP

WHEN YOU'RE SOURCING
MATERIALS FOR YOUR DESIGN,
IT HELPS TO THINK OUTSIDE
THE BOX. I'VE DISCOVERED
SOME GREAT PLANTER OPTIONS
IN THE KITCHEN SECTION
OF CHAIN STORES, SUCH
AS TERRACOTTA CROCKPOTS,
CERAMIC UTENSIL HOLDERS
AND GLASS JARS. ANOTHER
FABULOUS FIND WAS A VINTAGE
CERAMIC WATER FILTER.

PLANT PLACEMENT

The way you position plants – whether it's in a large outdoor garden or a small corner of a room – contributes to the overall feel of the space in many subtle ways. Here are a few tips for design-driven plant placement.

Work with odd numbers

Our eyes naturally find comfort in odd numbers, and we find solace in organic environments. When specifying plants and planters, work with odd numbers to make your gardens free-flowing. I have always been a big fan of asymmetry.

Create swathes

Grouping the same plant in long strips is an organised way to build your design. These swathes create a sense of abundance and movement within your garden, both indoors and out.

Dot the plants

Dotting your plants or feathering them off helps to make your garden more naturalistic and less rigid. It also softens the transitions between different plants. Traditionally, this technique has been used in outdoor gardens, but the same principle can be applied to indoor and balcony scenarios where planters are incorporated.

Rather than dotting the same plant species over a garden bed, you can dot the same plant across a larger cluster of planters. It may not have quite the same impact as large-scale in-ground planting, but this technique helps to unify loose planters so they feel like a family.

Bring the outside in by repeating the tones of exterior plants in the potted specimens that frame the view.

Run a plant thread

You can choose a single plant to become your thread in a garden. This might take the form of the same underplanting throughout the garden so that it becomes a carpet that unites the other plants. It's a particularly handy technique if you have a series of eclectic planters and plants that might not otherwise feel like a family.

Repeat types and tones

Repeating plant types and colours or planter tones is a great way to create calm in the garden. Our eyes naturally find comfort when they recognise a pattern in the environment.

Work to points

To give a garden focus, try working to points (see page 88). Select the specific areas you want to be focal points, and then allow the other elements to reveal themselves slowly.

PLAN A corner with three planters. The tallest and widest is in the corner with the other two snug next to it.

VIEW The same scene from the front to show height.

YOU CAN MAKE EVEN A SIMPLE CLUSTER FLOW ORGANICALLY BY LAYERING THE HEIGHTS.

Layering plants

To give your garden more dimension, layer the plants to add depth. Think about it as layers of clothing for your garden. By introducing a variety of different plants, you'll create a garden that has a deeper perspective and organic flow. Layering can be achieved in a large outdoor garden, but also within a single planter.

In an outdoor garden bed, you can draw inspiration from nature to layer a range of plants with different growth habits, much like we see in a natural ecosystem – forests comprise tall trees, under-canopy species and ground-level plants. A mix of trees, shrubs and ground covers, plus short, medium and large plants (see pages 54–64), will fill your garden with vertical layers of foliage.

In a potted scenario, your planters can also be layered. Place tall planters towards the back of your clusters, typically against walls or in

corners. Then, using planters with a range of heights, arrange them from tallest to shortest so the lowest planters are in the front. Once you have layered your planters, choose plants that assist with foliage layering. A mix of plant types with various growth habits creates an organic layering across the collection of planters. I particularly love underplanting within planters so foliage cascades over the edges. This softens the overall appearance of the cluster and also creates a sense of continuity when the same underplanting is used across a series of planters.

Even if you can fit only one planter into your space, layering can be done effectively. Often, we see a single plant type in a planter. However, by incorporating two or three plant types, you can create a small garden. Typically, I would pair a medium to tall plant with cascading underplanting.

CONNECTING ALL THE ELEMENTS

Curating a space is about fostering a seamless connection between all the elements. Here are some examples of design themes that have been brought together beautifully in a green space.

A BLACK MINIMALIST INTERIOR
This simple but bold aesthetic lends itself to strong silhouettes.

A COLOURFUL INTERIOR

Colour makes for a playful atmosphere where there is no limit to the tones you can mix together.

AN ECLECTIC PATINA LOOK

This collection of homely elements oozes nostalgia and familiarity.

A MEDITERRANEAN LOOK

Earthy and effortless, it celebrates rustic tones and patinas.

A WHITE SEASIDE LOOK

Channelling coastal tones,
the lighter palette provides
for an elevated look.

A WABI-SABI LOOK

Inspired by the ancient Japanese philosophy of accepting imperfection, it fosters an appreciation of objects as they are.

CHAPTER FOUR

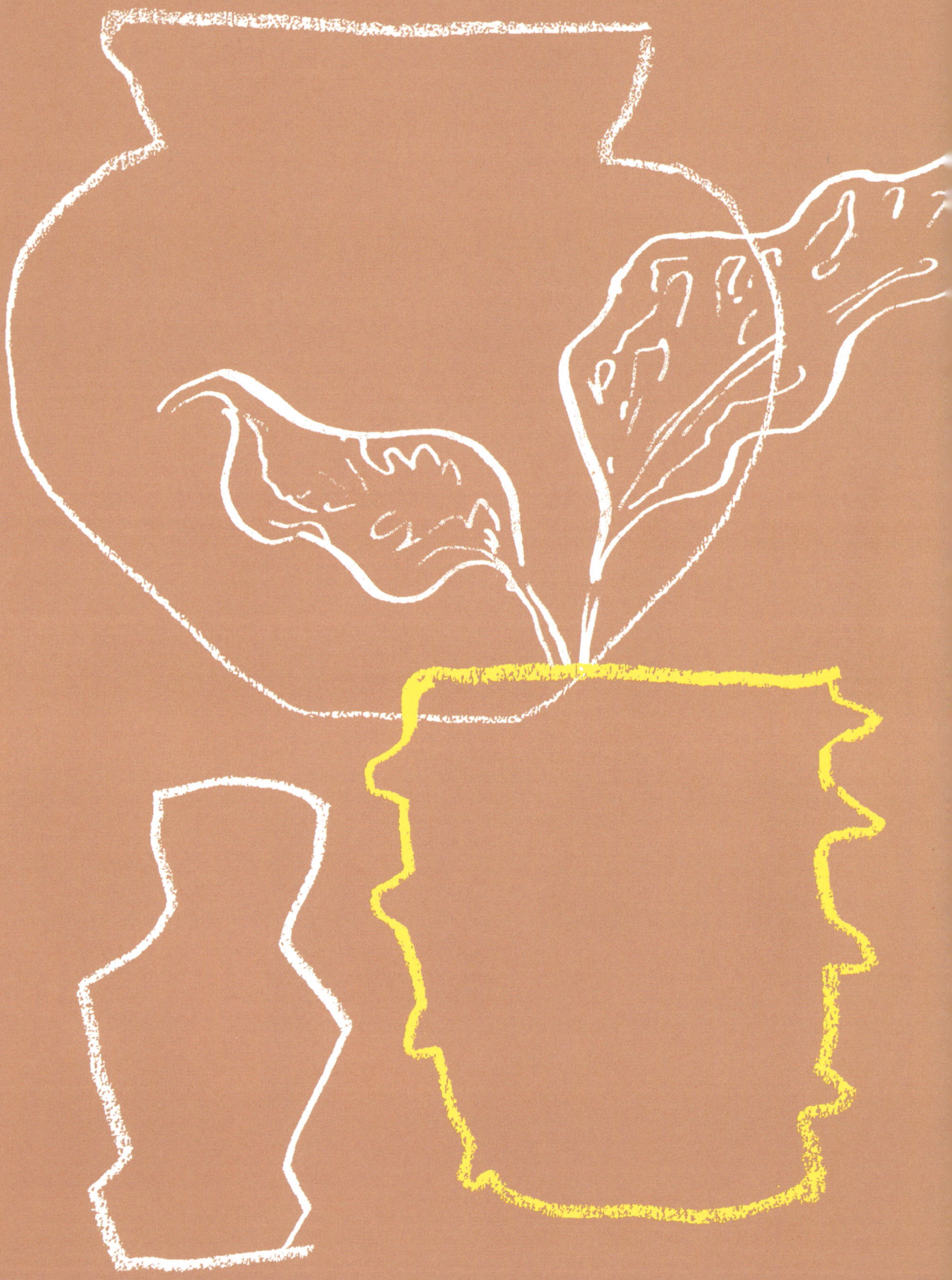

gardening techniques

I haven't always been at ease with gardening. What helped me was constantly reading, watching and listening to anything I could find on how to garden; exposing myself to new ideas and knowledge; speaking to other gardeners – and then just giving things a go. My gardening lessons have included successes – but also huge failures. The failures have taught me to try again, over and over.

The act of gardening shouldn't feel like a chore as, ultimately, gardening is life. And, throughout life, we need to be open to new experiences and knowledge. From this, we can form a skill set of gardening techniques that is developed over time and with practise.

TESTING, EDITING AND EXPERIMENTING

Whether it's honing a gardening skill or trying to finesse a style, gardeners are always learning and experimenting. So, treat the design and cultivation of your garden as a lifelong project. Find contentment in trialling your ideas or seeing your designs evolve as your garden matures. In a few years, you might decide that you want to change the colours, and this is completely fine. There might also be colour combinations you've never seen before and wish to try yourself. Go for it! That's the only way to find out if they'll be successful. If they don't work, then this adds to your knowledge bank.

The process of gardening is one of testing and editing, and you'll happily go around and around with new ideas and thoughts. Let curiosity guide you, and dare to try new things. In the meantime, I've compiled some useful gardening tips and techniques in this chapter to help you along the way, including transplanting, repotting, pruning, fertilising, mulching and propagating.

Key gardening tools

A well-rounded gardening kit will make tending to your garden a breeze. However, choosing the right range of tools will depend on the type of garden you have. If you're nurturing a container garden, then a small collection of hand tools may suffice.

Conversely, a large suburban garden will require an array of bigger tools, such as loppers, a spade and a crowbar. No matter what type of garden you have, ensure that you purchase top-quality gardening tools that will stand the test of time.

CROWBAR
SPADE
EDGER
GARDEN FORK
LOPPER
HOSE
WATERING-CAN

TOOLS

> CROWBAR
> EDGER
> GARDEN FORK
> HOSE
> LOPPER
> SPADE
> WATERING-CAN

ALSO USEFUL:

> FOLDING SAW
> GARDEN SAW
> HAND FORK
> HAND TROWEL
> HEDGER
> SECATEURS
> SNIPS
> SPRAYER
> TARP
> TRUG
> WEEDER
> WHEELBARROW
  OR BUCKET

Choosing a growing medium

The quality of your potting mix or soil is of the utmost importance when creating a garden. Which medium you use will depend on the plant types you choose to grow, as well as whether you're planting in a container or in the ground.

Potted plants

A general potting mix will cover most bases. To the right are recipes for two other mixes that I regularly use (multiply the amount to suit the size of your planter). However, if you prefer not to make your own, then garden centres and hardware stores offer a wide range of pre-mixed options to suit many plant species.

GENERAL POTTING MIX

This can be bought from your local garden centre or hardware store. Make sure that you purchase premium-grade potting mix.

GENERAL TROPICAL POTTING MIX

1 cup peat moss

1 cup premium-grade potting mix

1 cup perlite

2 tablespoons slow-release fertiliser

GENERAL ARID POTTING MIX

1 cup gravel

3 cups premium-grade potting mix

2 cups perlite

2 tablespoons slow-release fertiliser

Garden beds

Rather than introducing fresh soil to a garden bed, it's better to work with what you have and continuously build on the soil quality by adding organic matter in the form of compost, manure and organic minerals. However, if you choose to add garden soil because you don't have the time or space to gradually build up your soil quality, then make sure it's premium-quality garden soil. Don't use potting mix, as it's not suitable for in-ground use.

As I live in the country, I'm fortunate to have the space to create my own hot compost, where I process all of our household organic waste alongside immense amounts of garden waste, from grass clippings to everyday prunings. I've also lived in the inner city, where I didn't have the luxury of room to make homemade compost, and I found that ready-made compost from a garden centre or hardware store was just as beneficial.

To improve your existing soil, simply mix in layers of homemade or store-bought compost and water along with a seaweed solution to allow healthy microbes to settle in and multiply.

WHEN PREPARING GARDEN BEDS FOR PLANTING, INCORPORATE COMPOST AS A BASE TO REPLENISH YOUR SOIL AND BUILD A SOURCE OF GOOD MICROBES.

How to plant

Here is a step-by-step guide to moving plants from their original pots to new planters or the ground.

1   Gently press the sides of the pot to loosen the root ball.

2   Place the pot on its side.

3   While supporting the plant, gently slide it out of the pot.

4   Gently tease out the roots with your fingers, working around the circumference of the plant.

5   Loosen the base of the root ball to free the bottom of the root system. If the roots are wound tightly, then you may need to trim some of the root ball. I recommend trimming a maximum of one-third, so the plant doesn't go into shock.

6   **If planting into a planter** – Fill the bottom of the planter with a layer of potting mix, enough so that when the plant is placed into the planter, the crown (where the stem meets the roots) sits 1–2 centimetres ($\frac{1}{3}$–$\frac{1}{4}$ inches) lower than the rim of the planter.
If planting into the ground – Dig a hole approximately double the size of the root ball. Fill the bottom of the hole with a layer of soil, enough so that when the plant is placed into the hole, the crown sits level with the natural ground level.

7   If you wish, apply some organic or slow-release fertiliser to the base layer of potting mix or soil. Refer to the product packaging for application rates.

8   Place the plant onto the base layer of soil, and backfill the planter or hole with potting mix or soil. Make sure to gently compact the growing medium as you add it to remove any big air pockets. This will ensure that the root system makes good contact with the soil in order to access valuable nutrients.

9   Finish by watering in your plant.

SPACING YOUR PLANTS

Refer to the mature size and width of your plants when spacing out young plants, so they have enough room to grow happily without competing too much for light and nutrients. A quick online search will clarify this for you. However, I've found that you can plant a little tighter than is recommended for a denser, more abundant look. You can always prune or transplant as your garden establishes.

Repotting

When buying plants, choosing the correct size can be confusing. As a guide, the plant's root ball should be slightly smaller than your decorative planter. I recommend allowing a gap of approximately 3 centimetres (1 inch) for small planters, 6 centimetres (2⅓ inches) for medium planters and 10 centimetres (4 inches) for large planters between the circumference of the plant's grower's pot and the internal edge of the decorative planter. That way, you know that the plant will fit into the decorative planter before you try to repot it.

#2 Remove the plant from the pot by either tipping the plant on its side and gently pulling on the trunk, or gently holding the trunk and tapping off the pot. For rigid or heavy pots, it helps to place the pot on its side with a towel underneath. Gently pull the plant by the trunk; continue using the trowel to further prise the roots from the pot until the plant slides out. If you are working with bigger plants, you may need some help from a fellow gardener.

#1 Loosen the plant from its existing pot by pressing the sides of the pot. If the existing pot is made of a rigid material – such as stone or terracotta – you'll need to use a hand trowel or butterknife to loosen the root system. Push the trowel or knife down the inside of the pot edge and gently prise the root system away from the edge until the plant is free.

#3 Gently loosen the root system with a hand fork.

#4 Trim the bottom of the root ball by approximately one-third, aiming to remove any tangled roots.

#5 Place a layer of premium-grade potting mix into the bottom of the new planter, along with an appropriate amount of slow-release fertiliser (see the product packaging for details).

#6 Place the plant on top of this bed, and backfill the planter with more premium-grade potting mix so that the crown of the plant sits 1–2 centimetres (⅓–¼ inches) below the rim.

#7 Finish by watering in your plant.

What to do when a plant becomes too big

A thriving plant is a good problem to have. There are several ways to deal with a plant that has become too large for its home.

Prune excess branches

You can cut back the plant to the desired height and width. How much you prune will depend on your personal preference and your space. I like to prune a bit more than required so that when the new branches form, they have space to grow and won't quickly become too tall or too wide again. (For more pruning tips and techniques, see pages 124–9.)

Trim the roots of potted plants

One way to ensure that your growing plant has room to thrive is to use the root-pruning technique. This involves removing the plant from its planter, trimming the root system, and repotting it in the same planter (or a new planter if desired).

    To do this, follow the repotting steps on pages 114–15. After removing the plant from the planter and loosening the root system, trim up to one-fifth of the sides and bottom of the root ball. Once done, resume potting your plant.

Move large outdoor plants

Before digging up a plant from your garden and transplanting it to a more spacious location, note which way the plant is currently facing. You can mark the trunk or branches with a marker. If you place it back into the ground facing the same way, then the plant will have the best chance of settling into its new home. Now follow these steps:

1 Loosen the soil under the drip line of the plant using a hand fork. The drip line is an imaginary circle on the ground below the widest-reaching branches of the plant. It's where water drips from the plant and hits the soil's surface.

2 Once you've loosened the soil around the drip line, continue to gently prise up the plant, moving in a circle around the drip line.

3 If the plant is grounded firmly, then you may need to use a long crowbar or spade to put some pressure on the root ball. Insert the crowbar claw or spade blade under the loosened root ball, and put downward pressure on the tool's other end.

4 Continue doing this around the root ball until it has lifted from the ground.

5 Lie the plant on its side.

6 Trim the root ball back slightly, aiming to clean up any torn roots. Ensure that you don't prune too much from the root ball. As a guide, I would prune up to one-fifth of the root ball.

7 Dig a hole twice the size of the root ball in the plant's new position.

8 Backfill the hole slightly so that the crown of the plant will be sitting in the same position relative to the ground level as it was originally.

9 Place the plant into the hole, ensuring that the direction of the plant matches the original direction as marked.

10 Backfill the hole with soil, using your foot or hands to press down the soil as you go.

11 Water in well with a seaweed solution.

TRANSPLANTING

There are two main rules to bear in mind when transplanting.

1 Transplant when the weather is favourable for the plant. For instance, it's best to transplant deciduous plants in winter, while the ideal time to move tropical plants is in spring.

2 Avoid transplanting plants during periods of extreme heat or cold.

Feeding your plants

Feeding your plants is extremely important for healthy growth, and to maximise flowering and produce yield. I predominantly use organic soil additives, once every 6–12 months. The best times to feed your plants are during season changes, when planting into planters or when creating new garden beds. Additives come in many different forms, from compost to seaweed.

Compost

Adding compost matter to your garden helps to make the soil nutrient-rich, but compost takes some time to break down. This means that your plants receive nutrients over time, rather than instantly. Compost application therefore has longer-term, controlled benefits.

Liquid fertiliser

This feeding method provides balanced nutrients for your plants, and the liquid form allows the fertiliser to be taken up more quickly.

Granular fertiliser

Resembling grit, this type of fertiliser is typically formulated to break down over a short-term period and deliver important nutrients to your plants.

Manure

Composted manure has the benefit of improving the texture of the soil, creating a well-drained garden. It increases the number of beneficial microorganisms and breaks down slowly and steadily. Avoid using fresh manure because it can contain harmful bacteria. To be safe, purchase a pre-processed option from your local garden centre or hardware store.

Seaweed

Seaweed is jam-packed with nutrients that are beneficial in the garden. The easiest way to apply a seaweed solution is by using an over-the-counter concentrate. However, you can also make your own seaweed tea at home.

Gather washed-up seaweed from a beach. (Check with the local council first to make sure that it's okay to do so.) Collect enough seaweed to fill a lidded bucket or container.

Wash your foraged seaweed lightly, then place it into the bucket and top up with water. Put the lid on, but make sure that it isn't closed tightly. Place the bucket in a spot that's out of direct sunlight.

Stir your seaweed tea daily over the next week or so, then leave it to brew for approximately three months. After this time, dilute your mix to a ratio of one part seaweed tea to ten parts water, and use it on your garden plants.

Mulching

Mulch helps to retain soil moisture and minimises rapid temperature changes; it allows the soil to remain moist and cool in the warmer months while insulating it in the cooler months. Maintaining a good layer of mulch will also minimise the growth of weeds. Therefore, it's important to regularly add a thick layer of mulch to both your garden beds and your potted plants.

I use mulch only on outdoor plants because they're exposed to the biggest environmental shifts. Stick to using materials that break down over time, including woodchips (such as pine or eucalyptus), straw and shredded leaves. The choice of particle size is entirely personal, based on your desired look.

When mulching larger garden beds, try to source locally made mulch because it's the most affordable option – this allows you to adequately cover large areas without having to spend all of your gardening budget. Mulch bought in bulk typically comes in a coarser grain. The advantage is that it breaks down slowly, so it usually only needs to be applied once a year.

If you'd like a more delicate look, then use finely shredded mulch. It breaks down quickly so will be continuously improving soil health, but it will need to be replaced more often than coarser mulch.

As a rule of thumb for garden beds, apply a layer of mulch that is 5–10 centimetres (2–4 inches) thick. Too little mulch can reduce its effectiveness, and you'll need to reapply it more frequently. Too much mulch can lead to nutrient imbalances in the soil.

WOODCHIPS AND BARK

PEBBLES

SUGAR CANE

GRASS CLIPPINGS

HAY

STRAW

Types of mulch

There are many options when it comes to mulching materials. Each has its own pros and cons.

MULCH GUIDE

HAY

> This is a readily available material. It contains diverse plant matter – including clover, grasses, legumes and a range of flowers – and provides a rich variety of nutrients. Hay can be laid down on the soil relatively flat, and it breaks down rapidly. The downside of hay is that it usually contains grass seeds that can sprout.

STRAW

> Popular among vegie gardeners, straw is an affordable mulch to use in your garden. Unlike hay, straw is made from the stems of grain crops. Being a lightweight material, it is easy to work with, but the thick stems break down more slowly than hay, adding fewer nutrients to the soil. However, the stems are hollow and trap air, making them superb insulators and good at retaining moisture. Straw rarely contains grass or weed seeds.

WOODCHIPS AND BARK

> This is the most common mulch. It comes in a variety of timbers and particle sizes, and it can be fresh or composted. Fresh woodchip and bark mulch is typically less than three months old and is just starting to break down. A drawback of using fresh woodchip and bark mulch is that it tends to deprive your plants of nitrogen. On the other hand, when woodchips and bark have been composted for over three months, they are high in nutrients. Compared to straw, woodchips and bark create a better moisture barrier and last longer in your garden; they also provide a wider range of nutrients for the soil as they break down. Woodchip and bark mulch is widely used on ornamental garden beds and around trees.

SUGAR CANE

> Derived from the leaves and heads of sugar cane, this mulch is affordable and readily available. It's easy to work with and breaks down rapidly, promoting good soil microbes. It's widely used in vegie gardens.

GRASS CLIPPINGS

> If you have a lawn, then you'll have access to a lot of grass clippings. They're a simple and effective mulch that will add plenty of nutrients to the soil. Feel free to use fresh or dried clippings, but avoid adding grass clippings with seed heads to your garden.

PEBBLES AND GRAVEL

> Pebbles and gravel are the longest-lasting mulch type and assist in maintaining soil moisture; however, they don't improve soil health — you'll need to add nutrients manually. I recommend using gravel in arid gardens filled with succulents or Mediterranean species.

Building structures and support

Often, your plants need support when they're growing and maturing. Depending on your garden style, you may want the support to be a feature, or you may prefer it to be virtually invisible. When providing a support for a climbing rose, for example, you may choose to have a decorative structure to add an architectural form. This could be a vintage piece, a modern metal frame or lovely old branches sourced from your garden.

On the other hand, when correcting a plant that's growing in an undesired direction, you may prefer to use a simple bamboo or hardwood stake. Then you can make sure that it's concealed by the branch or trunk as much as possible.

There is a wide variety of materials you can use for support structures. I prefer natural materials unless I'm looking for longevity. If you'd like to use longer-lasting supports, choose materials such as metal and timber treated for outdoor use.

Bamboo stakes
Readily available, the humble bamboo stake is a versatile tool that can be used to straighten small- to medium-sized plants in your garden. It can also be used to support flower spikes.

Metal stakes
Whether they're brass, aluminium or powder-coated steel, metal stakes come in an array of simple and decorative forms. They are long lasting and can act as a sculptural addition to your garden.

Ceramic stakes
Handmade ceramic stakes are treasured pieces. They are less common in stores, so look out for talented ceramicists who create these beautiful accessories. Typically, they're ideal for small plants.

Branch prunings
We often forget that our gardens provide an array of items that can be repurposed. Prunings from trees or shrubs can make great plant supports. Branch prunings work especially well as support structures: stacked as a wall to create shelter, or formed into a trellis for climbing plants to scramble up.

Wire
Whenever I buy plants, they come with plant tags attached with wire. I make sure to collect these pieces of wire so I can re-use them in the garden. If you don't have a collection of recycled wire, then you can buy rolls of garden wire at your local garden centre or hardware store. You'll need a good-quality PVC-coated wire that can be easily spun around stakes, branches and stems. It's best to use a wire that's coated in PVC so it won't rust quickly in the elements or cut into fragile plant matter. Wire can be used to tie your plant's branches or trunk to a sturdy support, such as a stake or solid structure, to encourage the growth habit you desire. Make sure not to tie the wire too tightly; a little give will avoid damage to the plant as it grows. It's also a good idea to check on the wire every season to see if it needs to be loosened or refixed.

Cable wire
When it comes to creating a support for climbing plants, you want something that will last a lifetime. I like using tension wire and eyelets to run the cable through. To achieve this, the eyelets need to be fixed to a wall or onto a timber or metal pole. Another popular way to use cable wire is by fixing eyelets to verandah fascia boards and running the wire in front of the timber. You can easily make simple lines, or you can create an intricate pattern to add depth to your garden.

Twine
Natural twine is the ideal material to tie stray stems to supports or to anchor them to something more solid. Twine degrades faster than wire, so use it only as a short-term solution.

HARDWOOD STAKES

These are stronger than bamboo, allowing them to hold more weight and to withstand extreme wind. Use hardwood stakes in vegie gardens and for supporting larger plants and immature trees.

Pruning

Often, new gardeners are nervous about pruning. They feel that it might jeopardise their hard work and cause the plant to suffer. Actually, pruning is a positive way to manage and even promote growth in your garden. It's also a great way to encourage a specific look.

There are many reasons to prune plants. We often think about it as a method to control growth and remove unwanted branches. However, pruning can add a whole new perspective to your garden. Imagine it as a hairstyle change for your green space! Here are some of the scenarios in which you might need to prune your plants.

IT'S GOOD PRACTICE NOT TO USE INFECTED PRUNINGS AS MULCH OR PROPAGATION MATERIAL. YOU DON'T WANT TO INFECT HEALTHY PLANTS WITH A PEST OR DISEASE FROM ANOTHER PLANT.

Gangly growth

When plants become too tall or leggy (with long, straggly stems), pruning can help to promote a more compact and dense growth habit.

Dieback

Plants have their own defence system, which will take energy away from parts of the plant that are least important for survival. This leads to the deterioration of stems and branches. Pruning in this scenario helps to tidy up the plant and make room for new growth.

Pests and diseases

It's impossible to prevent pests and disease in the garden. As a gardener, you'll need to make a judgement regarding the best approach when treating your plants. In some cases, the pest or disease may have taken over only a certain part of your plant, so you can prune this affected section.

Topiary

Topiary is the art of creating interesting and decorative shapes in plant life through careful pruning. It allows you to fashion plants into desired forms, and can introduce balance and transition between the naturalistic and the designed elements in your garden.

Balance

All plants have unique natural qualities. As a gardener, you sometimes need to assist with balancing a plant that's lopsided or windswept. One way to do this is to prune the heavier side to even out the plant's weight.

Working with nature

When we want a plant to grow the way we imagine it, we can be quick to prune and shape to this perfect subconscious image. Having spent a lot of time in Japan, I've learned that there's another way to think about shaping and pruning: we can foster the way in which the plant or tree wants to grow in all its imbalance and uniqueness. We see this practice in bonsai, where gardeners let the plant reveal its personality. As gardeners, we should try to prune empathetically.

Flowers and produce

Pruning is a great way to encourage a plant to flower profusely or to produce a larger harvest. For example, the goal of pruning a fruit tree is to open up the tree, allowing more sunlight and air through, which increases fruit yield.

GREEN TIP
AS A GENERAL RULE,
ANY DISEASED OR
DEAD BRANCHES CAN
BE PRUNED AT
ANY TIME OF
THE YEAR.

When to prune

The best time to prune depends on the species. You can use a seasonal calendar to remind you when to prune the different plants in your garden. But since the seasons rarely follow a calendar precisely, there's a window of time during which you can extend your pruning schedule. Take note of the weather forecast to see if you have more time to complete your seasonal pruning.

Mid- to late winter
Prune deciduous plants and trees, such as apples (*Malus* spp.) and maples (*Acer* spp.).

Late winter to early spring
Prune and shape evergreen plants and trees, such as olive trees (*Olea europaea*), after the last frost.

Spring
Prune semi-woody perennials, such as butterfly bushes (*Buddleja* spp.), sages (*Salvia* spp.) and lavenders (*Lavandula* spp.); prune broadleaved evergreen plants, such as Oregon grape (*Berberis aquifolium*); prune deciduous shrubs, such as roses, smoke bushes (*Cotinus* spp.) and hydrangeas; trim back ornamental grasses, such as feather reed grass (*Calamagrostis × acutiflora*) and Chinese silver grass (*Miscanthus sinensis*); prune citrus trees after harvesting the fruits.

Early summer
Deadhead and lightly trim untidy branches from spring-flowering shrubs, such as butterfly bushes (*Buddleja* spp.).

Mid- to late summer
Prune evergreen plants, such as boxes (*Buxus* spp.) and other hedges, for additional shaping; trim untidy branches from sappy trees, such as silver birch (*Betula pendula*) and maples (*Acer* spp.).

Autumn
Deadhead perennials and annuals, such as yarrows (*Achillea* spp.) and fennel (*Foeniculum vulgare*).

How to prune

There are some fundamental pruning techniques that every gardener should have in their repertoire, regardless of whether they're caring for a small container garden or a large backyard. At first, these techniques may seem overwhelming. But they're easily mastered, and over time they'll become second nature to you.

Stem pruning

There are three main ways to prune stems: cutting off a single stem, cutting off a branch and pruning heavily. When cutting off a single stem, make the cut above a bud. Ensure that you cut diagonally away from the bud.

To cut off a branch, make the cut above a healthy side branch. Ensure that you cut diagonally away from the side branch.

The technique of pruning heavily is used to drastically decrease the bulk of a plant. It's typically performed in early spring or at the ideal time for the specific plant type. Choose a point on the plant below your desired height or width, and trim any thin branches with secateurs. You can cut back the plant more dramatically if you like, but make sure to leave 8–15 centimetres (3–6 inches) of plant matter above the crown to minimise the chance of your plant becoming stressed. For any thicker stems and branches, you'll need to use a garden saw to cut back the plant. Make sure to saw on an angle. This prevents water from sitting on the cut stem or branch and rotting it.

Tip pruning

Use either your fingers or fine snips to trim new growth on your plants.

How to prune
CONTINUED

Deadheading
This is the act of pruning away flowers that have faded. The simple act of cutting off spent flower heads often promotes another flush of flowers in both perennials and roses.

Pruning away deadwood
When branches are diseased or have died, you should prune them back as closely as possible to the nearest live branches.

Cutting back
This technique is widely used for tending to grasses and perennials. The process of cutting back a plant is typically done in early spring, when the plant is still dormant but about to sprout again. Plants are pruned to only a few centimetres from the ground.

Root pruning

Typically, this is performed on potted plants. After a few years of being in the same planter, plants will need more room to thrive. If you don't want to keep repotting into a bigger and bigger pot because of space constraints, then root pruning is a great way to keep your plants happy without needing a larger planter. The plant is removed carefully from the existing planter, and the root ball is pruned by up to one-fifth of the mass (see page 116). It's then replanted into the same planter with fresh premium-grade potting mix.

Pruning away cross branching

For deciduous plants, this is easiest to achieve in winter when the plants have lost their foliage, so you can clearly see the branch structure. Observe the way the branches are growing, and prune away any that cross inwards or that are growing astray.

Nurturing seeds and seedlings

Growing plants from seed is an affordable way to promote more variety in your garden. It also allows you to add seasonality through a range of flowering plants that are rarely sold as potted specimens. I love germinating plants such as poppies (*Papaver* spp.), cosmos, fennel (*Foeniculum vulgare*), amaranth and Queen Anne's lace (*Daucus carota*).

There are two main methods: sowing seeds into punnets, or sowing directly into garden beds. When preparing to sow seeds into punnets, make sure that the punnets are clean. Before sowing seeds into a garden bed, ensure that you remove any weeds and stones, then gently loosen the soil with a rake or hand fork.

Germinating in punnets

To raise seeds in punnets before transferring the seedlings to the garden, follow these steps.

1 Fill your seed punnets with fine seed-raising mix, and use a hard edge to scrape away the excess from the top. I like to use the edge of my hand trowel, but you can use a butterknife or even a stake.

2 Plant your seeds evenly into your punnets. Small or fine seeds can typically be sprinkled directly onto the soil, while larger seeds will need to be pressed into the soil and buried.

3 Once the seeds are planted, you can place a fine layer of vermiculite over the soil surface. This helps to maintain the moisture, which is especially useful for seed raising because seeds are not very forgiving if they dry out.

4 Make sure to label your seed punnets with the species or variety of plants you're raising.

When growing a range of different plants, it's easy to forget what you've planted. I tend to use plastic labels from the local hardware store, and write the names with a permanent marker. These can be re-used each season. You can also use old ice-cream sticks, cut-up pieces of hard plastic, or masking tape applied to the side of the punnets.

5 Water in your seeds with a gentle stream of water, making sure that the pressure doesn't wash away your seeds.

6 Place your punnets in a warm and bright position. To speed up the germination process, feel free to create a little greenhouse for your seeds by covering them with a plastic soft-drink bottle (with the bottom cut off and the lid still screwed on) or by placing them under a glass dome.

7 Ensure that the soil is kept moist while the seedlings develop.

8 As your seeds germinate and the seedlings grow, you may wish to feed them with a liquid fertiliser or blood and bone. Use the dosage recommended on the packaging.

Germinating in a garden bed

To sow seeds directly into a garden bed, follow these steps.

1   Boost the nutrient content of the existing soil by adding premium-quality garden soil or compost. Mix this into the top 10–15 centimetres (4–6 inches) using a rake or hand fork.

2   Plant your seeds into the garden bed, spacing them evenly. Refer to the seed packaging for the best time to plant your chosen species. Small or fine seeds can typically be sprinkled directly onto the soil, while larger seeds will need to be pressed into the soil and buried.

3   Once the seeds are planted, sprinkle a fine layer of soil over the seeds.

4   Water in your seeds with a gentle stream of water, making sure that the pressure doesn't wash away your seeds.

5   To prevent pests from digging up your seeds, cover the garden bed with shadecloth or bird netting. To do this, create some U-shaped frames out of rigid wire. Insert the two points of each frame into the soil, spacing the frames evenly and close enough to hold up the shadecloth or bird netting. Run the shadecloth or bird netting over the frames, making sure that you cover all access points to the seeds.

6   Ensure that the soil is kept moist while the seedlings develop.

7   As your seeds germinate and the seedlings grow, you may wish to feed them with a liquid fertiliser or blood and bone. Use the dosage recommended on the packaging.

#1 If your seeds are large and easy to handle, then use your finger to press each one into the soil.

#2 Once you've finished planting your seeds, cover them with a fine layer of soil.

Planting out seedlings

Once your punnet-grown seeds have established into seedlings, it's time to plant them into your garden or into planters.

1 Slowly prise the seedlings out of the punnets. A butterknife will help with this.

2 Gently loosen the soil around the roots so you can separate the seedlings into individual or smaller groups of plants.

3 Lay out your seedlings onto the soil in the locations you wish to plant them, so you can visually assess if they're spread out enough.

4 When you are happy with the spacing of your seedlings, plant them into the ground or planter so the soil is at the same level on the seedling stems as it was in the punnets. You can add some compost around the seedlings to give them extra nutrients and finish with a layer of fine mulch.

5 Water in your seedlings with a seaweed extract, which will help them to settle in faster.

#1 Carefully remove the seedlings from the punnet. Clear a space in the garden bed if necessary.

#2 Separate the seedlings into single plants.

#3 Plant each seedling, ensuring that they have enough space between them to grow.

Aerating

As your plants establish over time, the soil will become dense and compacted, both in planters and in the ground. Aerating your soil improves soil health by creating small spaces for water molecules and microorganisms. When provided with a good soil environment, your plant's root system will have access to a rich source of nutrients. The best time to aerate soil is during the plant's growing season.

To aerate your soil, simply use a large garden fork or small hand fork to penetrate the soil, and rock it gently back and forth. This can be done in garden beds as well as planters. For smaller planters, use a chopstick or bamboo stake to create narrow holes around each plant's root ball.

Training a plant

When a plant is growing in an undesired shape, you can help it to grow in a specific way by training it. There are a few ways this can be done.

Staking – For a lopsided plant that is top heavy, stake the plant to support it. (For more about staking, see page 122.)

Rotating – If your plant is in a planter and growing towards the light, then an easy way to balance the plant is to rotate your planter every month or so to promote even growth.

Wiring – A heavily practised technique in the art of bonsai, wiring plants is a great way to re-orientate young, malleable branches without having to prune and lose the growth. (For more information about wiring, see overleaf.)

How to wire a plant

You'll need tin snips and bonsai wire (anodised aluminium or annealed copper). The thickness of the wire will depend on the branch you want to shape; choose a thickness that will hold the branch in place. To wire the plant, follow these steps.

#2 Continue by winding the wire around the branch you want to bend.

#1 Start by winding the wire around the trunk at least twice for stability, at a 45-degree angle.

#3 Close to the end of the branch, wrap the wire back on itself to create a loop. Trim off any excess wire.

#4 Now that the branch is wired, gently manipulate the branch into the shape you're after. Be mindful not to bend the branch too far, otherwise it might break.

#5 You may need to gradually bend the branch over a few months to achieve the result you want.

Espalier

Small spaces require creative solutions. Espaliering ornamental and fruit trees is a great way to add height to your designs without taking up too much ground area. This age-old method lends itself to a wide range of patterns, allowing you to take a more graphic approach in your garden.

How to espalier a plant

You'll need a single tree or shrub (or multiple if you wish to create a wall), heavy-gauge wire, and screws or anchors to fasten the wire into masonry or timber.

1   Start by choosing the pattern for your espalier (for example, cordon, fan, candelabra or Belgian fence).

2   Measure the chosen location, and space out the trees or shrubs (if using more than one) so that you can visualise how they will grow. Tall elements work best against a wall towards the back of your space. This helps to provide depth and directs your gaze from front to back.

3   Prepare your support structure by anchoring a series of horizontal wires to a wall or posts using screws or a similar fixing material.

4   Once your support is set up, plant your tree(s) or shrub(s) around 30 centimetres (12 inches) from the support.

5   Train two branches to the bottom wire support by tying them to the wire with twine. As the plant grows, continue this process on each of the upper wires. Trim away any stray shoots as they develop.

ESPALIERING IN A PLANTER

If you're creating an espaliered plant in a planter, you can build a support structure that anchors into the planter. After potting your plant into the planter, push two strong stakes into the potting mix on either side of your plant. As a guide, allow 20 centimetres (8 inches) from the trunk on either side. Once the uprights are secured, you can fix additional stakes horizontally with either wire or twine to reinforce the structure. Then add horizontal wires between the two uprights.

Pest control

Pests are unavoidable in the garden. After all, we're working with nature, and animals and insects play an important role in the ecosystem. Nonetheless, there are ways you can reduce the chances of pests affecting your plants, or manage a pest outbreak when it occurs.

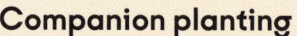

Natural pest control

Many insects bring benefits to our gardens, including pollination. So, the best approach to pest control is to minimise numbers rather than to eradicate the insects entirely. Although there are times when I need to use chemical-based pesticides, I prefer to tackle pests with natural and homemade remedies, such as cinnamon spray, vinegar spray, neem oil pesticide spray, and garlic and chilli spray (see opposite for recipes).

Physical prevention

It isn't always the most aesthetically pleasing approach, but incorporating a physical barrier can help to stop pests from reaching your plant. You can purchase interesting cloches (bell-shaped covers made from glass), or create your own with natural materials. Other handy forms of physical protection include:

> bird netting
> fine wire, such as chicken or aviary wire
> plastic guards
> bamboo cloches
> wire cages
> rope netting.

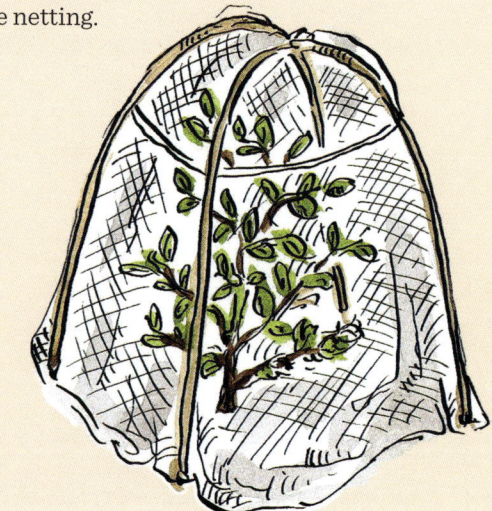

Companion planting

This is an organic way to use the properties of certain plants to naturally discourage pests from an area of your garden. It creates a partnership between the closely grown plants that benefits each type of plant. Companion planting is particularly popular among produce growers. Here are some examples of plant companions that pair well:

THYME (*Thymus* sp.)
and rose
The strong aroma of thyme foliage deters pests. When planted with rose bushes, it helps to mask the scented flowers that are much loved by pests.

MARIGOLD (*Tagetes* sp.)
and sage (*Salvia* sp.)
Both marigold and sage are strong attractors of pollinators, such as bees. Planting them together will increase flower production.

YARROW (*Achillea* sp.)
and vegetables
Yarrow naturally attracts beneficial pollinators and insects, while its scent deters aphids. Planting it near produce will increase pollination.

ROSEMARY (*Salvia rosmarinus*)
and dahlia
The fragrant scent of rosemary deters snails and slugs from feasting on the much-desired foliage of dahlia.

CHIVES (*Allium schoenoprasum*)
and mum (*Chrysanthemum* sp.)
The onion scent of chives deters aphids from feeding on mum flowers.

NATURAL PEST-CONTROL REMEDIES

CINNAMON SPRAY

To prevent and treat infestations of ants.

INGREDIENTS

2 teaspoons ground cinnamon

4 cups water

METHOD

Mix the ground cinnamon and water in a jar with the lid off, and let it sit overnight on your kitchen or laundry bench.

Strain the mixture using a fine sieve or fine cloth. Pour the strained liquid into a spray bottle.

HOW TO USE

> Mist the strained liquid onto any soil surface where ants are evident in large numbers or frequently seen.

NEEM OIL PESTICIDE SPRAY

To prevent and treat infestations of mealy bugs, thrips, whiteflies, aphids, mites and fungi.

INGREDIENTS

4 cups water

1 teaspoon dishwashing detergent

2 teaspoons neem oil

METHOD

In a spray bottle, mix the water, dishwashing detergent and neem oil.

HOW TO USE

> Spray on and underneath the foliage of your infested plants on a warm day, ensuring that all infected areas are covered in the solution. Avoid spraying in the early mornings and evenings when beneficial insects are active.

> Continue treatment until there are no traces of pests.

> Neem oil pesticide spray can also be used as a preventive measure on plants that are prone to pests. Simply spray on foliage and repeat throughout the growing season as required.

VINEGAR SPRAY

To prevent and treat infestations of various pests, including moths, slugs and ants.

INGREDIENTS

1 cup white vinegar

1 cup water

METHOD

In a spray bottle, mix the vinegar and water.

HOW TO USE

> Spray on and underneath the foliage of your infested plants, ensuring that all infected areas are covered in the solution.

> Continue treatment until there are no traces of pests.

GARLIC AND CHILLI SPRAY

To prevent and treat infestations of small sucking insects, including aphids, caterpillars and whiteflies.

INGREDIENTS

10 garlic cloves

8–10 hot chillies, or

2 tablespoons chilli powder

2 cups water

1 tablespoon dishwashing detergent

METHOD

Place the garlic, chillies or chilli powder and water into a blender. Blitz until smooth.

Place the mixture into a jar with the lid off, and let it sit overnight on your kitchen or laundry bench. Strain the mixture using a fine sieve or fine cloth.

Stir the dishwashing detergent into the strained liquid. Pour the liquid into a glass jar with a plastic lid.

HOW TO USE

> In a spray bottle, mix 2 tablespoons of the liquid with 1 litre (34 fluid ounces) of water.

> Spray the areas of your plants that are affected by an infestation.

> Make sure to also spray underneath foliage.

IT'S IMPORTANT NOT TO USE GARLIC AND CHILLI SPRAY ON YOUR PLANTS DURING EXTREMELY HOT WEATHER BECAUSE THE COMBINATION OF THE MIXTURE AND THE HEAT MAY BURN YOUR PLANTS.

Seasonality

Caring for a seasonal garden seems like a chore to some people, as they believe that this type of garden requires more maintenance than others. However, there are many benefits to a seasonal garden. If we see gardening as therapy, then raking up leaves or deadheading spent flowers is calming; the changes in the garden throughout the year create an atmosphere that shifts during the seasons.

    I always make sure to add a range of flowering and/or deciduous plants outdoors, even if they're subtle additions.

My favourite flowers include:
- avens (*Geum* spp.)
- citruses
- creeping thyme (*Thymus serpyllum*)
- dahlias
- lavenders (*Lavandula* spp.)
- orchid cacti (*Epiphyllum* spp.)
- poppies (*Papaver* spp.)
- rosemary (*Salvia rosmarinus*)
- sages (*Salvia* spp.)
- wax plants (*Hoya* spp.)

My favourite deciduous species are:
- Chinese wisteria (*Wisteria sinensis*)
- crab apple (*Malus sylvestris*)
- crepe myrtles (*Lagerstroemia* spp.)
- eastern redbud (*Cercis canadensis*)
- English oak (*Quercus robur*)
- maidenhair tree (*Ginkgo biloba*)
- maples (*Acer* spp.)
- persimmon (*Diospyros kaki*)
- quince (*Cydonia oblonga*)
- saucer magnolia (*Magnolia × soulangeana*)

WHAT TO DO AS THE SEASONS CHANGE

Gardening throughout the seasons doesn't have to be daunting. Some tasks (such as deadheading flowers) aren't that important in the grand scheme of things, and they can be done when there's time. If you're late with performing certain jobs, this can actually add a naturalistic feel to your garden. Here's a quick guide to the tasks you should plan to do throughout the seasons.

SPRING
> Apply a fresh layer of mulch.
> Apply nutrients to the soil.
> Plant any new plants.
> Make sure to weed early in the season.

SUMMER
> Perform a light prune of any plants that need slight shaping.
> Apply a liquid seaweed solution to any plants stressed by the heat.
> Continue weeding.

AUTUMN
> Apply a fresh layer of mulch where your mulch has broken down quickly.
> Apply fertiliser to the soil in late autumn so your plants can access nutrients before becoming dormant.

WINTER
> Prune dormant plants around mid- to late winter.
> Plant any new plants, after any chance of frost has passed.

Plant propagation

Propagating plants is a simple and rewarding way to expand your collection without having to buy large quantities to populate your garden. It's also a great way to share plants with friends and family, or to source some rarer plant species that you've spotted in a friend's garden. In addition, you can propagate plants that have worked well in your garden to fill bare patches in your garden beds or to extend into other spaces within your home.

There are a few key propagation techniques that will form the basis of your gardening skill set, such as stem cutting, leaf cutting, leaf sectioning and division. With these easy methods, you'll be able to propagate plants throughout the seasons.

STEM CUTTING

This technique works well for:
- butterfly bushes (*Buddleja* spp.)
- fiddle leaf figs (*Ficus lyrata*)
- fruit salad plants (*Monstera deliciosa*)
- roses
- rubber plants (*Ficus elastica*)
- sages (*Salvia* spp.).

Materials
secateurs
rooting hormone (optional)
propagating pots or trays
soil or propagating mix

Method
1 Using your secateurs, trim a branch from a plant you want to propagate. Ensure that you make a straight cut just below a node (leaf joint). Your cutting should have two or three nodes.

2 Using your secateurs again, remove two or three leaves from the bottom of the cutting.

3 Dip the bottom of the cutting in rooting hormone (if using), and place it into the soil or propagating mix. You can also place the cutting directly into water, and allow it to strike roots.

4 Place your stem cutting in a bright position that receives filtered light.

5 Once the cutting has a healthy root system and is showing signs of new growth, you can plant it into the ground or a planter.

LEAF CUTTING

This technique works well for:
- African violets (*Streptocarpus ionanthus*)
- begonias
- donkey's tails (*Sedum morganianum*)
- jade plants (*Crassula ovata*)
- radiator plants (*Peperomia* spp.)
- snake plants (*Dracaena* spp.).

Materials
secateurs or scissors
rooting hormone (optional)
propagating pots or trays
soil or propagating mix

Method

1 Using secateurs or scissors, trim off a healthy leaf close to the stem.

2 Dip the cut end into rooting hormone (if using).

3 Place the cutting into soil or propagating mix, with the tip of the leaf above the surface.

4 Place your leaf cutting in a bright position that receives filtered light. To speed up propagation, place a mini greenhouse over the cutting.

5 Once the cutting has a healthy root system and is showing signs of new growth, you can plant it into the ground or a planter.

LEAF-PETIOLE CUTTING

This technique works well for:
- African violets (*Streptocarpus ionanthus*)
- chestnut vines (*Tetrastigma voinierianum*)
- radiator plants (*Peperomia* spp.)
- wax plants (*Hoya* spp.)
- Zanzibar gems (*Zamioculcas zamiifolia*).

Materials
secateurs or scissors
propagating pots or trays
soil or propagating mix

Method
1 Take a cutting by trimming a leaf off the main plant at the base of the petiole (leaf stalk), near the soil's surface.

2 Place the cut end into soil or propagating mix.

3 Place your leaf-petiole cutting in a bright position with filtered light to allow the plant to root.

4 Once the cutting has a healthy root system and is showing signs of new growth, you can plant it into the ground or a planter.

LEAF-BUD CUTTING

This technique works well for:
• *Cissus* 'Ellen Danica'
• devil's ivy (*Epipremnum aureum*)
• hydrangeas
• raspberry (*Rubus idaeus*)
• rubber plants (*Ficus elastica*).

Materials
secateurs
propagating pots or trays
soil or propagating mix

Method

1 Using your secateurs, take a cutting from your main plant by trimming a piece of the branch or stem between nodes (leaf joints). Make sure that it is a straight cut, and keep the leaf attached. You require only one node, but you can propagate with more. Propagating with more nodes on the one cutting will create a more established plant.

2 Plant the cutting so that the stem sits horizontally under the soil or propagating mix. Allow the remaining leaf or leaves to sit above the surface.

3 Place your leaf-bud cutting in a bright position that receives filtered light.

4 Once the cutting has a healthy root system and is showing signs of new growth, you can plant it into the ground or a planter.

LEAF SECTIONING

This technique works well for:
- African violets (*Streptocarpus ionanthus*)
- gloxinias (*Sinningia speciosa*)
- painted-leaf begonias (*Begonia rex*).

Materials
secateurs or scissors
sharp knife
propagating pots or trays
soil or propagating mix

Method

1 Select a large, healthy leaf from the main plant.

2 Remove it at the petiole (leaf stalk) using secateurs or scissors.

3 Make wedge-shaped cuts through the leaf with a sharp knife, ensuring that each wedge contains a central vein.

4 Place the leaf section wedges into soil or propagating mix that is approximately 2–5 centimetres (¾–2 inches) deep.

5 Place your leaf sections in a bright position in filtered light.

6 Once the leaf sections have a healthy root system and are showing signs of new growth, you can plant them into the ground or a planter.

149

SUCKERS AND RUNNERS

This technique works well for:
- bamboos (*Bambusa* spp.)
- mints (*Mentha* spp.)
- spider plants (*Chlorophytum comosum*)
- strawberry (*Fragaria × ananassa*).

Materials
sharp knife or secateurs
propagating pots or trays
soil or propagating mix

Method

1 Remove smaller offset plants from the mother plant by cutting the sucker or runner at the base of the main plant. Suckers or runners should only be removed once they've developed some mature leaves, with a few evident roots at the base.

2 Plant your suckers or runners in soil or propagating mix.

3 Place your suckers or runners somewhere that receives filtered light, but keep them sheltered from extreme sunlight and wind.

4 Once the suckers or runners have a healthy root system and are showing signs of new growth, you can plant them into the ground or a planter.

DIVISION

This technique works well for:
- bearded iris (*Iris × germanica*)
- blue star ferns (*Phlebodium aureum*)
- *Calamagrostis × acutiflora* 'Karl Foerster'
- cast iron plants (*Aspidistra elatior*)
- Chinese silver grasses (*Miscanthus sinensis*)
- peace lilies (*Spathiphyllum* spp.).

Materials
sharp knife or trowel
pots (if using)
soil or propagating mix

Method

1 Remove the plant you are dividing from its pot or carefully lift it out of the ground.

2 Divide the large original plant into smaller plants by using a sharp knife or trowel to cut vertically up through the roots and the base of the plant several times.

3 Plant the divided plants into pots as per the repotting tips on pages 114–15, or directly into the garden.

4 Place the potted plants in an area that receives filtered light, and protect them from harsh winds and direct sunlight. If planting into your garden, then ensure that the lighting conditions are similar to those found in the plant's original location.

Mail-order plants

Online shopping opens the plant world to all gardeners. No longer are we limited to the nurseries that we can physically visit. Instead, a treasure trove of new species and varieties is available at our fingertips. Ordering plants online is an easy process, but there are a few things you need to know so you can settle in your plants on arrival.

Plants are typically sent in one of three ways:

1 Bare-rooted

This means that the grower's pot and the soil are both removed. The root ball is wrapped in a wet medium (such as newspaper) and then wrapped in a plastic bag to preserve moisture during transit.

2 Grower's pot removed

The soil is kept on the root ball and is wrapped in either a plastic bag or newspaper to preserve moisture during transit.

3 Potted plant

The plant is sent with its pot intact, and the pot is wrapped in a plastic bag to preserve moisture during transit.

Often, plants are cut back or pruned to fit them into a box and to limit the plant's stress level. Typically, plants are shipped when they are dormant for optimal transitioning. However, they can also be shipped during their growth period.

After receiving your online plants:

> Make sure to open the packaging as soon as possible.

> Avoid exposing the plants to extreme heat or cold.

> Unpack each plant gently. Placing plants that have been shipped without a pot into a plastic tray with sides helps to keep them upright. Remove all packaging to allow the plant to breathe.

> For plants that have been bare-rooted, it helps to soak the roots in a seaweed solution overnight to calm the plants. Once soaked, plant them into planters or straight into your garden.

> For plants where the pot has been removed, plant them into planters or straight into your garden (if you have a spot for them already).

> Water in with a seaweed solution to calm your plants.

garden design recipes

I've always enjoyed being inspired by dishes that chefs create and pass on through their recipes. In cooking, there's a vast array of styles and levels of difficulty, and recipes are a straightforward way of sharing the craft. Here, I've adapted this concept to make designing and looking after gardens more accessible to more people.

HOW TO READ A
GARDEN DESIGN 'RECIPE'

These easy-to-follow recipes provide inspiration and ideas for simple and effective ways to add greenery to your home, inside and out. In each recipe, the ideas are tailored to a particular space, to create an atmosphere and enhance the overall design aesthetic.

     If you fall in love with a recipe but want to use it in a space that's different from the one pictured, then feel free to adapt it. If your space has different lighting conditions, for example, then you can adapt the planting to suit. Observe the foliage colours, textures and growth habits, and use these characteristics to source plants that are better suited to your space.

OPTIMAL LIGHTING CONDITIONS
This is a guide to the lighting range that best suits the plants in the recipe.

TIP
This extra information will help with styling and looking after your plants.

WATERING
Advice on how much to water your plants throughout the various seasons.

THE PLANT SOCIETY DESIGN HANDBOOK

WHEN WORKING WITH A LARGE-SCALE DECORATIVE PLANTER, POT YOUR PLANT INTO A LIGHTER GROWER'S PLANTER FIRST AND THEN PLACE THIS INTO THE HEAVIER PLANTER. YOU CAN THEN EASILY LIFT OUT THE LIGHTER PLANTER TO WASH THE FOLIAGE DOWN IN THE BATHROOM OR OUTDOORS.

Table canopy

When curating plants around tables, we often play it safe and lean towards small tabletop gestures. But why not make a bigger and more exciting moment of the humble dining table? After all, it's the centre of any home. One time, I was yearning for a different mood for a home I was working in, and I thought about how nice it would be to eat and work under a tree canopy. So, I found a tree and gently trained it into the required shape. (To learn how to train a tree, see page 138.)

Filtered to bright light

KIT OF PARTS

PLANTS
1 kauri pine (*Agathis robusta*)
3 devil's ivies (*Epipremnum aureum*)

PLANTER
1 neutral-coloured planter, approx. 40 cm diameter and 60 cm high

PLANTING
premium-quality indoor potting mix

PREPARATION
Ensure that your planter has a drip tray to prevent your floor from becoming stained.

METHOD
The kauri pine you select for this design will become a sculptural element, so take your time when considering the tree's form and trunk shape. It doesn't need to be perfect. Choose a tree that's a little gnarly and askew.

Place a layer of potting mix into the planter, approximately 20–30 centimetres deep, and place the kauri pine.

Backfill with potting mix, leaving enough room to plant the devil's ivies evenly around the kauri pine. Add the devil's ivies, then continue to backfill with potting mix, making sure to gently press down to compact the potting mix.

Position the planter against your table. If possible, let the tabletop overlap the planter a little. It may seem unconventional, but this haphazardness will add to the atmosphere of your space.

Finish by watering in your plants.

170

GARDEN DESIGN RECIPES

WATERING
In the warmer months, keep the potting mix moist but not drenched.

In the cooler months, water less; allow the potting mix to dry out for approximately one week between watering.

171

KIT OF PARTS
These are the 'ingredients' you need for your garden project, including plants, planters and potting material.

METHOD
A step-by-step guide to creating your project. (For advice on planting, see pages 112–15.)

Earthy cluster

Heavy materials (such as marble and granite) are very popular in kitchen design. You can soften the look and feel of stone benchtops with a simple, considered grouping of plants. Here, I drew inspiration from the rich walnut joinery and soft tones in the marble. To create a seamless look, I carried the grey tones into two of the planters while embracing a rich, earthy ochre planter as a feature in the front of the grouping. It's important to play with differences in height across the three planters so there's an organic flow.

 Filtered to bright light

KIT OF PARTS

PLANTS

1 peace lily (*Spathiphyllum* sp.)
1 *Cissus* 'Ellen Danica'
1 mistletoe cactus (*Rhipsalis* sp.)

PLANTERS

1 mid-grey planter, approx.
   22 cm (8½ in) diameter
1 light grey planter, approx.
   18 cm (7 in) diameter
1 burnt orange or russet planter,
   approx. 25 cm (10 in) diameter

PLANTING

premium-quality indoor
   potting mix

PREPARATION

Ensure that your planters have drip trays to prevent your benchtop from becoming stained.

METHOD

Choose a place in your kitchen for your cluster of plants, preferably a surface that you don't use so you won't need to move the plants. The front corners of an island bench or a corner of the main bench (such as beside the cooktop or next to a tall pantry) are ideal.

Plant the peace lily into the mid-grey planter, *Cissus* 'Ellen Danica' into the light grey planter and the mistletoe cactus into the burnt orange or russet planter. Water in your plants.

Place the peace lily towards the back of the nominated area. This provides some height and becomes a backdrop for the other plants.

Place *Cissus* 'Ellen Danica' to the left of the peace lily. It helps if you offset the two planters slightly, so they are not both adjacent to the straight lines of the benchtop.

Place the mistletoe cactus towards the front, almost creating a triangle with the other planters when looking down on the bench. Use a shallow planter towards the front, so it allows you to glimpse the other planters.

WATERING

In the warmer months, keep the potting mix moist but not drenched.

In the cooler months, water less; allow the potting mix to dry out for approximately one week between watering.

Desktop landscape

In modest spaces, I adore extremely considered moments. A curated focal point of greenery can bring joy to work and study sessions. However, small additions don't always need to be simple. They can hold immense layers of detail and intricacy, even within a single planter.

 Filtered to bright light

KIT OF PARTS

PLANTS

1 queen of hearts (*Homalomena* sp.)

2 fine-leaved wax plants (*Hoya* spp.)

2 handfuls woodland moss – this can often be bought at florists

PLANTER

1 low ceramic planter, approx. 25 cm (10 in) diameter and 10 cm (4 in) high

PLANTING

premium-quality indoor potting mix

PREPARATION

Ensure that your planter has a drip tray to prevent your desk from becoming stained.

METHOD

Fill your planter to approximately one-quarter of its height with potting mix (this may vary based on the size of your plants). Place the queen of hearts into the planter, off-centre. Place a wax plant on either side of the queen of hearts to create an organic flow.

Backfill the planter, and gently press down to compact the potting mix and remove any large air pockets. Leave a 1-centimetre ($\frac{1}{3}$-inch) gap between the potting-mix surface and the top of the planter.

Evenly place a layer of moss on the potting-mix surface and under the foliage to finish your miniature landscape. This will act as mulch for your little ecosystem.

Finish by watering in your plants. Refresh the moss as needed.

WATERING

In the warmer months, keep the potting mix moist but not drenched.

In the cooler months, water less; allow the potting mix to dry out for approximately one week between watering.

AS YOU'RE PLANTING THE QUEEN OF HEARTS AND WAX PLANTS INTO A SHALLOW PLANTER, YOU MAY NEED TO TRIM SOME EXTRA OFF THE ROOT BALL OF YOUR PLANTS. I RECOMMEND THAT YOU TRIM NO MORE THAN A THIRD OF THE ROOT BALL.

Verdant plant shelves

Plant shelves are the ideal way to soften a wall and instantly add character to an area. Space is often at a premium, both indoors and out, so plant shelves allow you to add layers to harsh architectural elements or bland walls. I'm a huge fan of using cascading plants, such as wax plants (*Hoya* spp.), *Cissus* 'Ellen Danica', devil's ivy (*Epipremnum aureum*), mistletoe cacti (*Rhipsalis* spp.), Boston ivy (*Parthenocissus tricuspidata*), cascading forms of rosemary (*Salvia rosmarinus*), *Dichondra argentea* 'Silver Falls' and string of bananas (*Curio radicans*).

 Filtered to bright light

KIT OF PARTS

PLANTS

2 *Cissus* 'Ellen Danica'
2 fine-leaved wax plants (*Hoya* spp.)
3 heartleaf philodendrons
　(*Philodendron hederaceum*)

PLANTERS

2 medium assorted white planters,
　approx. 20 cm (8 in) diameter
5 small assorted white planters,
　approx. 15 cm (6 in) diameter

PLANTING

premium-quality indoor
　potting mix

PREPARATION

Ensure that your planters have drip trays to prevent your shelves from becoming stained.

METHOD

Pot the two *Cissus* 'Ellen Danica' plants into the two medium planters.

Pot the wax plants and heartleaf philodendrons into the five small planters. Once planted, water in all of your plants.

Use the two *Cissus* 'Ellen Danica' plants to anchor your shelves by placing them diagonally across from each other. Next, place the heartleaf philodendrons under the higher *Cissus* 'Ellen Danica' in a triangular cluster.

Place a wax plant next to the lower *Cissus* 'Ellen Danica' and the heartleaf philodendrons. This helps to feather off your greenery.

Feel free to style other objects (such as books, ceramics and artwork) within your shelves to layer your interior with atmosphere.

WATERING

In the warmer months, keep the potting mix moist but not drenched.

In the cooler months, water less; allow the potting mix to dry out for approximately one week between watering.

For a quick touch of greenery, simply place one *Cissus* 'Ellen Danica' on a high shelf, and one heartleaf philodendron on a low shelf, slightly offset from the *Cissus*.

IF YOU MAKE SURE THAT YOUR PLANTERS ARE ALL WHITE, THEN THIS ALLOWS FOR A SEAMLESS CONNECTION. THE PLANTERS DON'T NEED TO BE THE SAME SHAPE OR FORM.

Indoor hanging garden

Why not bring greenery inside to create a lush cocoon? Hanging gardens take advantage of unused vertical space, freeing up the floor area below for everyday living. Incorporate a range of natural materials (such as cane and rattan) with matt neutral tones to make indoor spaces more sophisticated. When clustering your hanging planters into groups, work in odd numbers. For this project, we'll hang a group of three and a group of five.

 Filtered to bright light

KIT OF PARTS

PLANTS

3 devil's ivies (*Epipremnum aureum*)

2 wax plants (*Hoya* spp.)

1 orchid cactus (*Epiphyllum* sp.)

1 chain of hearts (*Ceropegia linearis* subsp. *woodii*)

1 monkey mask plant (*Monstera adansonii*)

PLANTERS

3 medium assorted hanging rattan/cane planters, 22–30 cm (8½–12 in) diameter

3 self-watering planters, to fit inside your rattan or cane planters

5 small white hanging planters, approx. 15 cm (6 in) diameter

PLANTING

premium-quality indoor potting mix

PREPARATION

For this project, you'll need a sturdy ceiling support (such as ceiling beams or hooks installed into the ceiling). It's important that your support has structural integrity, so you may want to consult a builder.

METHOD

When it comes to matching your plants to your planters, mix up the combinations for a free-flowing and organic result. Simply pair the plants with your preferred planters, and assess how they look in combination before repotting them.

Hang the planters in corners away from areas you may walk through. Start by hanging a medium rattan/cane planter in the corner.

Next, select a small white planter and hang it slightly in front of and to the side of the rattan/cane planter.

Continue hanging the remaining planters at different heights and distances from each other, ideally in one group of five and one group of three.

Once you're happy with the layout of your hanging garden, make sure to water in your plants.

WATERING

In the warmer months, keep the potting mix moist but not drenched.

In the cooler months, water less; allow the potting mix to dry out for approximately one week between watering.

Even a group of five hanging planters will provide visual interest, especially if they're hung at different heights.

TO MAKE WATERING EASIER WHEN USING RATTAN/CANE PLANTERS INDOORS, POT THE PLANTS INTO SELF-WATERING PLANTERS FIRST. THEN PLACE THE SELF-WATERING PLANTERS INTO THE MORE DECORATIVE RATTAN/CANE PLANTERS.

165

Eclectic console

Consoles are the windows to our personalities and daily rituals. We use them to house our favourite objects and daily necessities. An eclectic mix of ceramic planters on a console can create a mood that oozes character. Pair the planters with cute and zany succulents and cacti for a uniquely sculptural design.

 Bright to harsh light

KIT OF PARTS

PLANTS

1 African milk tree (*Euphorbia trigona*)

1 pencil cactus (*Euphorbia tirucalli*)

1 elephant's foot (*Dioscorea elephantipes*)

1 Brazilian edelweiss (*Sinningia leucotricha*)

1 Hottentot bread (*Fockea edulis*)

3 pebble plants (*Lithops francisci*)

PLANTERS

1 large planter in an earthy tone, approx. 20 cm (8 in) diameter

1 medium planter in a saffron tone, approx. 15 cm (6 in) diameter

2 small planters in a burnt orange tone, approx. 10–12 cm (4–4¾ in) diameter

2 small off-white planters, approx. 10 cm (4 in) diameter

2 extra-small off-white planters approx. 8 cm (3 in) diameter

PLANTING

premium-quality arid potting mix

fine gravel (earthy tones)

PREPARATION

Ensure that your planters have drip trays to prevent your console from becoming stained.

METHOD

Pot the African milk tree into the large planter. This will be the focal point in the design.

Next, pot the pencil cactus into the medium planter, the elephant's foot and the Brazilian edelweiss into the two small burnt orange planters, and the Hottentot bread and a pebble plant into the two small off-white planters. The remaining pebble plants can then be planted into the two extra-small planters. Dress your planters with a layer of fine gravel, and lightly water in your plants.

When curating your planters on your console, begin with your tallest specimen (in this case, the large planter). Place this a little off-centre. It helps to mentally divide your console into halves.

Next, place your medium planter to the left of the large planter. One small burnt orange planter will then go to the right and slightly in front of the large planter. Place the small off-white planters on either side of the medium planter.

Place the other small burnt orange planter on the other side of the console, leaving a breathing gap to allow the planters to have their own space to shine.

To finish, place the two extra-small planters close to the lone small burnt orange planter.

WATERING

In the warmer months, keep the potting mix moist but not drenched.

In the cooler months, water less; allow the potting mix to dry out for approximately one week between watering.

WHEN REPOTTING
SPIKY OR THORNY PLANTS,
IT HELPS TO HAVE SOME
PADDING AROUND THE PLANT
TO PROTECT YOUR HANDS.
I LIKE TO RE-USE BUBBLE
WRAP OR FOAM FROM OLD
RETAIL PACKAGING.

Bold floor cluster

Monochromatic interiors are timeless. It's an aesthetic that creates a bold statement through its simplicity. I often design with charcoal or white planter tones so the planters create simple silhouettes in the background and allow the plants to stand out.

 Filtered to bright light

KIT OF PARTS

PLANTS

1 kauri pine (*Agathis robusta*)

6 *Cissus* 'Ellen Danica'

2 Chinese money plants (*Pilea peperomioides*)

PLANTERS

1 extra-large charcoal planter, approx. 50 cm (20 in) diameter and 50 cm (20 in) high

1 large charcoal planter, approx. 40 cm (16 in) diameter and 40 cm (16 in) high

1 medium charcoal planter, approx. 30 cm (12 in) diameter and 30 cm (12 in) high

PLANTING

premium-quality indoor potting mix

PREPARATION

Ensure that your planters have drip trays to prevent your floor from becoming stained.

METHOD

Place a layer of potting mix into the extra-large planter, approximately 20–30 centimetres (8–12 inches) deep. Position the kauri pine in the centre of the planter.

Backfill with potting mix, leaving enough room to plant three *Cissus* 'Ellen Danica' plants evenly around the circumference of the kauri pine. Add the three *Cissus*, then continue to backfill with potting mix, making sure to gently press down to compact the potting mix.

Place a layer of potting mix into the large planter, approximately 25–30 centimetres (10–12 inches) deep. Position the two Chinese money plants and two *Cissus* 'Ellen Danica' plants in the planter. (I like to lay out plants organically. In this instance, I'd plant the Chinese money plants in one half of the planter, and the *Cissus* 'Ellen Danica' plants in the other half.) Backfill with potting mix, making sure to gently press down to compact the potting mix.

Plant the last *Cissus* 'Ellen Danica' plant into the medium planter.

To style, position the extra-large planter against a wall, with a gap of approximately 5–10 centimetres (2–4 inches) between the planter and the wall. Position the large planter to the left but slightly overlapping the extra-large planter, leaving a gap of 20 centimetres (8 inches) between the planters. Position the medium planter to the right and approximately 5–10 centimetres (2–4 inches) in front of the extra-large planter.

Finish by watering in your plants.

WATERING

In the warmer months, keep the potting mix moist but not drenched.

In the cooler months, water less; allow the potting mix to dry out for approximately one week between watering.

WHEN WORKING WITH A
LARGE-SCALE DECORATIVE
PLANTER, POT YOUR PLANT
INTO A LIGHTER GROWER'S
PLANTER FIRST AND THEN
PLACE THIS INTO THE HEAVIER
PLANTER. YOU CAN THEN EASILY
LIFT OUT THE LIGHTER PLANTER
TO WASH THE FOLIAGE DOWN
IN THE BATHROOM
OR OUTDOORS.

Table canopy

When curating plants around tables, we often play it safe and lean towards small tabletop gestures. But why not make a bigger and more exciting moment of the humble dining table? After all, it's the centre of any home. On a recent project, I was looking for a different mood, and I thought about how nice it would be to eat and work under a tree canopy. So, I found a tree and gently trained it into the required shape. (To learn how to train a tree, see page 135.)

 Filtered to bright light

KIT OF PARTS

PLANTS

1 kauri pine (*Agathis robusta*)

3 devil's ivies (*Epipremnum aureum*)

PLANTER

1 neutral-coloured planter, approx. 40 cm (16 in) diameter and 60 cm (24 in) high

PLANTING

premium-quality indoor potting mix

PREPARATION

Ensure that your planter has a drip tray to prevent your floor from becoming stained.

METHOD

The kauri pine you select for this design will become a sculptural element, so take your time when considering the tree's form and trunk shape. It doesn't need to be perfect. Choose a tree that's a little gnarly and askew.

Place a layer of potting mix into the planter, approximately 20–30 centimetres (8–12 inches) deep, and place the kauri pine.

Backfill with potting mix, leaving enough room to plant the devil's ivies evenly around the kauri pine. Add the devil's ivies, then continue to backfill with potting mix, making sure to gently press down to compact the potting mix.

Position the planter against your table. If possible, let the tabletop overlap the planter a little. It may seem unconventional, but this haphazardness will add to the atmosphere of your space.

Finish by watering in your plants.

WATERING
In the warmer months, keep the potting mix moist but not drenched.

In the cooler months, water less; allow the potting mix to dry out for approximately one week between watering.

Bedroom intimates

SOURCE A RUBBER PLANT THAT IS A LITTLE GNARLY AND NATURALISTIC TO ADD CHARACTER TO YOUR CLUSTER. BE INSPIRED BY THE ART OF BONSAI FOR THIS DESIGN, AND MANICURE YOUR PLANTS TO BECOME STATEMENTS.

Bedrooms are not the most obvious spaces to create gardens, but greenery is immensely beneficial for our respite. Plants also help to purify the air we breathe. Given the amount of time we spend in the bedroom every day, I couldn't imagine a better environment in which to introduce a pair of curated planters. We look towards holidays as our escape, but with the right styling, the bedroom can become more than just a room to sleep in.

 Filtered to harsh light

KIT OF PARTS

PLANTS

1 rubber plant (*Ficus elastica*)

2 maidenhair vines (*Muehlenbeckia complexa*)

1 snake plant (*Dracaena* sp.) – I prefer cylindrical snake plant (*Dracaena angolensis*), but you're welcome to use one of your preference

PLANTERS

1 grey glazed planter, approx. 20 cm (8 in) diameter and 25 cm (10 in) high

1 blue-toned planter, approx. 15 cm (6 in) diameter and 15 cm (6 in) high

PLANTING

premium-quality indoor potting mix

PREPARATION

Ensure that your planters have drip trays to prevent your furniture from becoming stained.

METHOD

Place a layer of potting mix into the grey planter, approximately 5 centimetres (2 inches) deep, and place the rubber plant.

Backfill with potting mix, leaving enough room to plant the two maidenhair vines either side of the rubber plant. Add the maidenhair vines, then continue to backfill with potting mix, making sure to gently press down to compact the potting mix.

Plant the snake plant in the blue-toned planter.

When curating your cluster, leave enough room on the bedside table directly beside the bed so you can still place objects for everyday use. Style the rubber plant on the right of your bedside table, offset from the wall by approximately 2 centimetres ($^3/_4$ inch).

Place the snake plant to the left of the rubber plant, overlapping it slightly.

Finish by watering in your plants.

WATERING

In the warmer months, keep the potting mix moist but not drenched.

In the cooler months, water less; allow the potting mix to dry out for approximately one week between watering.

Desk trinkets

The world of succulents and cacti is filled with quirky forms and wonderfully sculptural characters. This collection of zany arid plants will become almost pet-like, and will keep you company while you're working away.

 Bright to harsh light

KIT OF PARTS

PLANTS

1 Madagascar ocotillo (*Alluaudia procera*)

1 old man cactus (*Cephalocereus senilis*)

1 Medusa's head (*Euphorbia caput-medusae*)

1 Peruvian torch cactus (*Trichocereus macrogonus*)

1 mistletoe cactus (*Rhipsalis* sp.)

PLANTERS

5 rustic planters in neutral tones (such as brown and off-white), 10–15 cm (4–6 in) diameter

PLANTING

premium-quality arid potting mix

fine gravel

PREPARATION

Ensure that your planters have drip trays to prevent your desk from becoming stained.

METHOD

Pot up each plant into a separate planter, and cover the surface of the potting mix with gravel.

To cluster your plants, create two groupings on your desk. On the right-hand side, start by placing the Madagascar ocotillo to the back of your desk.

Next, place the old man cactus and Medusa's head in front of the Madagascar ocotillo.

On the left-hand side of your desk, place the Peruvian torch cactus to the back and then the mistletoe cactus in front.

WATERING

In the warmer months, water lightly and allow the potting mix to dry out between watering.

In the cooler months, water sparingly and keep mostly dry.

STYLING TIP
THE PLANTERS CAN BE ECLECTIC IN STYLE, WITH CUTE FEET, SPIKY FORMS OR ROUGH TEXTURES.

174

WHEN AESTHETICALLY GOING OVER THE TOP, KEEP AN ELEMENT CONSISTENT TO ENSURE THAT THERE'S A THREAD RUNNING THROUGH THE DESIGN. HERE, WE'RE USING THE NATURAL TONES OF THE PLANTERS TO TIE EVERYTHING TOGETHER.

Bold balconies

Balconies are spaces that we often simply look out onto or use occasionally for respite. They come in a wide range of sizes and shapes, from small covered areas to large exposed spaces. Usually tiled or featuring a concrete finish, they can appear void of life. Bold and architectural plants in simple silhouettes can make strong statements on balconies. Curating a simple balcony garden with repetition of plant and planter types will help to draw a stronger connection between indoors and outdoors.

 Partial shade to full sun

KIT OF PARTS

PLANTS

2 rubber plants (*Ficus elastica*)

6 kidney weeds (*Dichondra repens*)

PLANTERS

1 narrow and tall charcoal planter, approx. 20 cm (8 in) diameter and 40 cm (16 in) high

1 wide and short charcoal planter, approx. 30 cm (12 in) diameter and 20 cm (8 in) high

PLANTING

premium-quality indoor potting mix

METHOD

Place a layer of potting mix into the tall planter, approximately 20–35 centimetres (8–14 inches) deep, and place one rubber plant.

Backfill with potting mix, leaving enough room to plant three kidney weeds evenly around the rubber plant. Their placement doesn't need to be perfect because they will eventually spread out and cover the surface as they grow. Continue to backfill with potting mix, making sure to gently press down to compact the potting mix.

Plant the short planter with the same combination and technique.

Position the tall planter in the corner of your balcony. Position the short planter to the side, slightly offset from the tall planter.

Finish by watering in your plants.

WATERING

In the warmer months, keep the potting mix moist but not drenched. During summer, you may need to water once or twice a day for your plants to thrive.

In the cooler months, water less; allow the potting mix to dry out for a couple of days between watering.

FOR LARGER BALCONIES, YOU CAN EASILY MULTIPLY THE NUMBER OF PLANTERS. HOWEVER, IT'S BEST TO WORK WITH ODD NUMBERS TO CREATE AN ORGANIC FLOW.

If you're designing for a small balcony, then you may not have room for two large planters. Try three small planters of different heights and widths instead.

Silver outdoor tones

When designing gardens, I love seeing how different foliage tones can create different moods. Silver-toned foliage has a distinctive elegance to it. The lighter shade channels Mediterranean, arid and coastal environments. Silver tones are used more and more in the modern garden as a way of creating a light and airy atmosphere.

 Partial sun to full sun

KIT OF PARTS

PLANTS

1 mountain cabbage tree (*Cussonia paniculata*)

6 *Cissus* 'Ellen Danica'

1 foxtail agave (*Agave attenuata*)

1 fan aloe (*Kumara plicatilis*)

PLANTERS

1 large rough oceanic planter, approx. 40 cm (16 in) diameter and 50 cm (20 in) high

1 medium rough oceanic planter, approx. 40 cm (16 in) diameter and 40 cm (16 in) high

1 small rough oceanic planter, approx. 25 cm (10 in) diameter and 35 cm (14 in) high

2 extra-small rough oceanic planters, approx. 20 cm (8 in) diameter and 20 cm (8 in) high

PLANTING

premium-quality arid potting mix

METHOD

Place a layer of potting mix into the large planter, approximately 20–30 centimetres (8–12 inches) deep, and place the mountain cabbage tree.

Backfill with potting mix, leaving enough room to plant two *Cissus* 'Ellen Danica' plants evenly around the mountain cabbage tree. Add the two *Cissus*, then continue to backfill with potting mix, making sure to gently press down to compact the potting mix.

In the medium planter, plant the foxtail agave with two *Cissus* 'Ellen Danica' plants around it.

For the small planter, plant the fan aloe by itself.

Finally, plant two *Cissus* 'Ellen Danica' plants into each of the extra-small planters.

Begin curating your cluster by placing the large planter in the corner of your space. To the left of this, place the medium planter. Then place the small planter to the right of the large planter.

In front of this cluster, place the two extra-small planters. Leave a slight gap between the back three planters and the front two planters.

Finish by watering in your plants.

WATERING

In the warmer months, keep the potting mix moist but not drenched. During summer, you may need to water once or twice a day for your plants to thrive.

In the cooler months, water less; allow the potting mix to dry out for a couple of days between watering.

OCEANIC PLANTERS ARE HIGHLY TEXTURED, MUCH LIKE OYSTER SHELLS. FEEL FREE TO REPLACE THEM WITH ANY TEXTURED PLANTERS OF YOUR CHOICE.

Olive tree hero

The olive tree (*Olea europaea*) is beloved worldwide for its ornamental grace. With its hardy Mediterranean bloodline, it's ideal for bright and harsh outdoor conditions. Either plant a series of trees to create some rhythm in your garden, or – as we've done here – feature a sculptural potted centrepiece on a balcony, deck or front porch.

 Partial sun to full sun

KIT OF PARTS

PLANTS

1 olive tree (*Olea europaea*)

5 cascading rosemaries (*Salvia rosmarinus*)

PLANTER

1 extra-large terracotta planter, approx. 65 cm (26 in) diameter and 70 cm (28 in) high

PLANTING

lightweight stones, such as scoria

geotextile fabric

premium-quality general-purpose potting mix

METHOD

Position the planter in the desired location. Fill the bottom quarter of the planter with lightweight stones, such as scoria. This will assist with providing drainage.

Cut a square of geotextile fabric to place over the stones. When placing the geotextile fabric into the planter, fold up the edges so they hug the internal walls of the planter.

Fill the planter with potting mix, stopping when there's still enough room for the root ball of the olive tree. Place the olive tree into the centre of the planter.

Backfill with potting mix, leaving enough room to plant the cascading rosemaries evenly around the olive tree. Continue to backfill with potting mix, making sure to gently press down to compact the potting mix.

Finish by watering in your plants.

WATERING

In the warmer months, keep the potting mix moist but not drenched. During summer, you may need to water once or twice a day for your plants to thrive.

In the cooler months, water less; allow the potting mix to dry out for approximately two weeks between watering.

IT'S BEST TO
HAVE TWO GARDENERS
FOR THIS PROJECT.
BECAUSE OF THE SIZE AND
WEIGHT OF THE PLANTER,
IT'S IMPORTANT TO HAVE
AN EXTRA SET OF HANDS
TO HELP MOVE IT.

181

Edible cluster

Citrus trees are a favourite for growing in planters in a range of urban and suburban scenarios. Their glossy foliage adds a strong textural form along paths and on balconies, the flowers offer a sweet aroma, and the ripe fruits can be eaten straight from the tree. Here, we're grouping together edible plants so they're easier to care for. I've chosen herbs that I use all the time, but feel free to replace them with herbs that suit your kitchen needs.

 Partial sun to full sun

KIT OF PARTS

PLANTS

1 citrus tree of your choice – I like using lemons, limes and cumquats, but any citrus will work

3 creeping thymes (*Thymus serpyllum*)

1 sage (*Salvia officinalis*)

1 mint (*Mentha* sp.)

PLANTERS

1 extra-large terracotta planter, approx. 50 cm (20 in) diameter and 60 cm (24 in) high

1 medium charcoal planter, approx. 40 cm (16 in) diameter and 30 cm (12 in) high

1 small terracotta planter, approx. 30 cm (12 in) diameter and 30 cm (12 in) high

PLANTING

lightweight stones, such as scoria

geotextile fabric

premium-quality general-purpose potting mix

METHOD

Position the extra-large planter in the desired location. Fill the bottom quarter of the planter with lightweight stones, such as scoria. This will assist with providing drainage.

Cut a square of geotextile fabric to place over the stones. When placing the geotextile fabric into the planter, fold up the edges so they hug the internal walls of the planter.

Fill the extra-large planter with potting mix, stopping when there's still enough room for the root ball of the citrus tree. Place the citrus tree into the centre of the planter.

Backfill with potting mix, leaving enough room to plant the three creeping thymes evenly around the citrus tree. Continue to backfill with potting mix, making sure to gently press down to compact the potting mix.

Place the medium planter to the left of the extra-large planter, fill with potting mix, and plant the sage. I don't use scoria in smaller planters because there's less risk of the planters becoming waterlogged. Backfill, gently pressing down to compact as you go.

Place the small planter to the right of the extra-large planter, fill with potting mix, and plant the mint.

Finish by watering in your plants.

WATERING

In the warmer months, keep the potting mix moist but not drenched. During summer, you may need to water once or twice a day for your plants to thrive.

In the cooler months, water less; allow the potting mix to dry out for approximately one week between watering.

183

Entry symmetry

An entry to a space is like a gardener's handshake with their guests. A lush moment can make a great first impression, but it's also a giving gesture to your local community. As your plants grow and sprawl out, they bring life to what would otherwise have been a hard surface.

 Partial sun to full sun

KIT OF PARTS

PLANTS

2 *Acer palmatum* 'Dissectum'

1 *Casuarina glauca* 'Cousin It'

8 *Dichondra argentea* 'Silver Falls'

PLANTERS

2 large white planters, approx. 60 cm (24 in) diameter and 50 cm (20 in) high

PLANTING

premium-quality general-purpose potting mix

METHOD

Position your planters either side of your front door.

Place a layer of potting mix into the planters, approximately 20–40 centimetres (8–16 inches) deep, and place one *Acer palmatum* 'Dissectum' into each planter. Centre the trunks.

Backfill with potting mix, leaving enough room to plant *Casuarina glauca* 'Cousin It' and *Dichondra argentea* 'Silver Falls'.

Plant *Casuarina glauca* 'Cousin It' into one of the planters. I like to make sure that my designs are not too symmetrical and feel a little organic.

Place three *Dichondra argentea* 'Silver Falls' plants evenly around the same planter, taking into consideration *Casuarina glauca* 'Cousin It'.

In the second planter, place the five remaining *Dichondra argentea* 'Silver Falls' plants evenly around *Acer palmatum* 'Dissectum'.

Continue to backfill both planters with potting mix, making sure to gently press down to compact the potting mix.

Finish by watering in your plants.

WATERING

In the warmer months, keep the potting mix moist but not drenched. During summer, you may need to water once or twice a day for your plants to thrive.

In the cooler months, water less; allow the potting mix to dry out for approximately one week between watering.

Outdoor table cluster

Outdoor seating arrangements can quickly feel more lived in with the addition of low table planting. As they'll be exposed to the weather – including wind and drought – the chosen plants need to be hardy and suited to the ever-changing environment. Select plants that will still look great with minimal cultivation.

- Partial sun to full sun

KIT OF PARTS

PLANTS

3 two-row stonecrops (*Phedimus spurius*)

1 prickly pear (*Opuntia* sp.)

1 *Cotyledon orbiculata* 'Silver Waves'

PLANTERS

1 medium terracotta planter, approx. 20 cm (8 in) diameter and 30 cm (12 in) high

1 small terracotta planter, approx. 15 cm (6 in) diameter and 10 cm (4 in) high

1 small saffron-toned planter, approx. 15 cm (6 in) diameter and 10 cm (4 in) high

PLANTING

premium-quality arid potting mix

small- to fine-grade gravel

METHOD

Place some potting mix into the medium planter. Plant the three two-row stonecrops evenly around this planter. Backfill with potting mix, making sure to gently press down to compact the potting mix.

Place some potting mix into the small terracotta planter. Plant the prickly pear into this planter. Backfill with potting mix, making sure to gently press down to compact the potting mix.

Place some potting mix into the small saffron-toned planter. Plant the *Cotyledon orbiculata* 'Silver Waves' plant into this planter. Backfill with potting mix, making sure to gently press down to compact the potting mix.

Dress all of the planters with a layer of gravel approximately 1 centimetre ($\frac{1}{3}$ inch) thick.

Place the medium planter in the centre of your table. Next, place the two small planters clustered together on the left-hand side of the medium planter.

Finish by watering in your plants.

WATERING

In the warmer months, water thoroughly; allow the potting mix to dry out between watering. During summer, you may need to water once a fortnight for your plants to thrive; however, if you choose not to water regularly, your plants will survive.

In the cooler months, water sparingly and keep mostly dry. You may not need to water at all if there's good rainfall.

Climbing wall

Sadly, with increased urban development, there's sometimes little room between our walls and those of our neighbours, and these tight spaces can be extremely uninspiring. Thankfully, we can turn to plants and gardens to bring a natural touch to the narrow walkways between buildings. I would rather look out onto greenery than a blank wall.

 Partial shade to partial sun

KIT OF PARTS

PLANTS

3 creeping figs (*Ficus pumila*)

2 Boston ivies (*Parthenocissus tricuspidata*)

6 native violets (*Viola hederacea*)

PLANTING

a garden bed that is at least 30 cm (12 in) wide

compost

fine-grade mulch

PREPARATION

These plants will cover a 2-metre (6-foot)-long wall. If your wall is longer, then make sure that you have enough plants to repeat the pattern.

METHOD

Prepare your garden bed by turning the soil and adding compost (see page 118).

Evenly space out the three creeping figs in a straight line against the wall. Between your creeping figs, evenly space the two Boston ivies, also against the wall. Now place the six native violets on the outer side of the creeping figs as well as between the creeping figs and Boston ivies.

Once you're happy with the spacing, start planting at one end and work your way to the other.

When the planting is finished, add a layer of mulch approximately 5 centimetres (2 inches) thick.

Finish by watering in your plants.

WATERING

In the warmer months, keep the soil moist but not drenched. During summer, you may need to water once or twice a day for your plants to thrive.

In the cooler months, water less; allow the soil to dry out for approximately one week between watering.

From little things big things grow; it won't take long for the young creeping figs and Boston ivies to race up your wall and fill the space with greenery.

Hanging trio for a verandah

It's hard to walk past a lush, overflowing verandah and not fall head over heels in love with the outpouring of foliage. However, don't be envious of fellow gardeners with hundreds of plants on their porch. It's so easy to begin, and I must warn you that it becomes an addiction. Start your journey with a trio of hanging planters.

 Partial sun to full sun

KIT OF PARTS

PLANTS

1 monkey tail cactus (*Cleistocactus winteri* subsp. *colademono*)

1 mistletoe cactus (*Rhipsalis* sp.)

1 zigzag cactus (*Disocactus anguliger*)

PLANTERS

1 large white hanging planter, approx. 30 cm (12 in) diameter and 16 cm (6⅓ in) deep

1 medium white hanging planter, approx. 20 cm (8 in) diameter and 15 cm (6 in) deep

1 medium green hanging planter, approx. 20 cm (8 in) diameter and 15 cm (6 in) deep

PLANTING

premium-quality arid potting mix

PREPARATION

You may need to thread wires for hanging as per the product description for your chosen planters.

METHOD

Place a layer of potting mix into the large white planter, approximately 10–15 centimetres (4–6 inches) deep, and place the monkey tail cactus in the centre. Backfill with potting mix, making sure to gently press down to compact the potting mix.

In the medium white planter, plant the mistletoe cactus. Then plant the zigzag cactus into the medium green planter. Backfill both with potting mix, making sure to gently press down to compact the potting mix.

Hang the large white planter approximately 1.6 metres (5¼ feet) from the floor, at the edge of the verandah. Make sure that it's in a location where you won't bump into it.

Hang the medium white planter to the left of the large white planter. Stagger these in height: hang the medium white planter around 15 centimetres (6 inches) higher than the large one. Also ensure that the medium white planter is approximately 5 centimetres (2 inches) in front of the large white planter.

Hang the medium green planter to the right of the other two, slightly lower than the medium white planter.

Finish by watering in your plants.

WATERING

In the warmer months, keep the potting mix moist but not drenched. During summer, you may need to water once a day for your plants to thrive.

In the cooler months, water less; allow the potting mix to dry out for two to three weeks between watering.

FEEL FREE TO CHOOSE A COLOUR OTHER THAN GREEN FOR THE THIRD PLANTER. YOU CAN MATCH OR CONTRAST WITH OTHER COLOURS IN YOUR GARDEN DESIGN.

Bathroom respite

Bathrooms can quickly become humid environments, which makes them ideal spaces to create a little microclimate in our homes. A green nest of foliage will create a gentle and relaxing environment in which to shower or bathe – turning these everyday activities into daily self-care routines.

 Filtered to bright light

KIT OF PARTS

PLANTS

1 mistletoe cactus (*Rhipsalis* sp.)

1 rabbit's foot fern (*Davallia solida* var. *fejeensis*)

1 air plant (*Tillandsia* sp.)

PLANTERS

1 small charcoal hanging planter, approx. 10 cm (4 in) diameter and 10 cm (4 in) high

1 small grey planter, approx. 15 cm (6 in) diameter and 10 cm (4 in) high

1 small rattan tray or bowl

PLANTING

premium-quality indoor potting mix

PREPARATION

You'll need to fix a hook into the ceiling for the hanging planter (see below for the correct placement).

METHOD

Place a layer of potting mix into the small charcoal hanging planter, approximately 5 centimetres (2 inches) deep, and place the mistletoe cactus. Backfill with potting mix, making sure to gently press down to compact the potting mix.

Place a layer of potting mix into the small grey planter, approximately 2–5 centimetres ($^3/_4$–2 inches) deep, and place the rabbit's foot fern. Backfill with potting mix, making sure to gently press down to compact the potting mix.

Position the small charcoal hanging planter so that it hangs beside your vanity. On the vanity, place the small grey planter closer to the sink. Beside this, place the air plant on top of the rattan tray or bowl.

Finish by watering in your potted plants.

WATERING

In the warmer months, keep the potting mix moist but not drenched.

In the cooler months, water less; allow the potting mix to dry out for approximately one week between watering.

Green library

IN THIS DESIGN, WE'RE USING THE SAME COLOUR FOR ALL OF THE PLANTERS. THIS UNIFIES THE DESIGN AND ENSURES THAT THE PLANTS ACT LIKE THE BACKUP DANCERS TO YOUR LIBRARY. KEEPING THE DESIGN SIMPLE MAKES THE OTHER ITEMS IN YOUR LIFE STAND OUT.

Open shelving is commonplace in most homes nowadays, as it's an ideal way to display all of the things we love. We're finding ways to decorate our homes with more and more layers, from books to sculptures and photos. Incorporating greenery buffers in your life library can quickly and easily make your space feel less cluttered and more considered.

 Filtered to bright light

KIT OF PARTS

PLANTS

5 heartleaf philodendrons (*Philodendron hederaceum*)

3 wax plants (*Hoya* spp.)

PLANTERS

8 small to medium white planters, approx.12–20 cm (4¾–8 in) diameter and 12–20 cm (4¾–8 in) high

PLANTING

premium-quality indoor potting mix

PREPARATION

Ensure that your planters have drip trays to prevent your shelves from becoming stained.

METHOD

Pot each plant into a separate planter.

Begin styling by placing a cluster of three heartleaf philodendrons and two wax plants. Place these in alternating positions to create the effect of a green wall.

Next, find space on another shelf to cluster two heartleaf philodendrons and one wax plant.

Finish by watering in your plants.

WATERING

In the warmer months, keep the potting mix moist but not drenched.

In the cooler months, water less; allow the potting mix to dry out for approximately one week between watering.

If you don't have room for eight planters on your shelves, then use fewer. You'll find that just three heartleaf philodendrons and two wax plants will still provide useful greenery.

Lush reading nook

Even within our own homes, we need a space in which to wind down and feed our inner introvert. Curating that special space for our wellbeing is a rewarding process. When I imagine this space, it's somewhere I can feel calm and comforted by nature – and there are no limits to how many plants or what style to use. I've always swayed towards natural materials (such as timber and stone) that develop a patina, as I love a home that shows the marks of time and the people living there.

 Filtered to bright light

KIT OF PARTS

PLANTS

1 weeping fig (*Ficus benjamina*)

7 *Epipremnum aureum* 'Marble Queen'

2 fruit salad plants (*Monstera deliciosa*)

2 giant sword ferns (*Nephrolepis biserrata*)

PLANTERS

1 extra-large white planter, approx. 50 cm (20 in) diameter and 50 cm (20 in) high

2 large white planters, approx. 40 cm (16 in) diameter and 40 cm (16 in) high

2 medium white planters, approx. 30 cm (12 in) diameter and 30 cm (12 in) high

2 small white planters, approx. 20 cm (8 in) diameter and 20 cm (8 in) high

PLANTING

premium-quality indoor potting mix

PREPARATION

Ensure that your planters have drip trays to prevent your floor from becoming stained.

METHOD

Place a layer of potting mix into the extra-large planter, approximately 10–20 centimetres (4–8 inches) deep, and place the weeping fig. Place three *Epipremnum aureum* 'Marble Queen' plants evenly around the main plant. Backfill with potting mix, making sure to gently press down to compact the potting mix.

Plant a fruit salad plant into each of the large planters. Backfill with potting mix, leaving enough room to plant one *Epipremnum aureum* 'Marble Queen' in each planter, then backfill around the two *Epipremnum*.

In each of the medium planters, plant one giant sword fern. Then plant one *Epipremnum aureum* 'Marble Queen' into each of the two small planters.

To style, place the extra-large planter in the corner behind your chair. Next to this, place one of the large planters so it's slightly tucked away behind the seat. Place the other large planter to the left of the chair.

Place one medium planter next to the extra-large planter; the other one will sit to the left of the chair, next to the large planter. Then position the two small planters on either side of the chair.

Finish by watering in your plants.

WATERING

In the warmer months, keep the potting mix moist but not drenched.

In the cooler months, water less; allow the potting mix to dry out for one to two weeks between watering.

Kitchen fronds

Kitchens with a strict focus on storage can become very utilitarian. I like using overhead shelves to draw attention to the heart of the home, giving it a different perspective: rather than a place for cooking alone, it's a vital space for conversation.

 Filtered to bright light

KIT OF PARTS

PLANTS

1 *Cissus* 'Ellen Danica'

1 mistletoe cactus (*Rhipsalis* sp.)

1 chain of hearts (*Ceropegia linearis* subsp. *woodii*)

PLANTERS

1 small but tall blue planter, approx. 15 cm (6 in) diameter and 20 cm (8 in) high

1 small blue planter, approx. 15 cm (6 in) diameter and 15 cm (6 in) high

1 small grey planter, approx. 15 cm (6 in) diameter and 15 cm (6 in) high

PLANTING

premium-quality indoor potting mix

PREPARATION

Ensure that your planters have drip trays to prevent your joinery from becoming stained.

METHOD

Place a layer of potting mix into the small but tall blue planter, approximately 10–15 centimetres (4–6 inches) deep, and place the *Cissus* 'Ellen Danica' plant. Backfill with potting mix, making sure to gently press down to compact the potting mix.

Plant the mistletoe cactus into the small blue planter. Then plant the chain of hearts into the small grey planter.

To style your plants, position *Cissus* 'Ellen Danica' on the shelf, offsetting it from the front of the shelf by approximately 5 centimetres (2 inches). Place the chain of hearts to the right so that it sits 1 centimetre ($\frac{1}{3}$ inch) from the front of the shelf. Lastly, position the mistletoe cactus so that it's slightly tucked in behind the chain of hearts.

Finish by watering in your plants.

WATERING

In the warmer months, keep the potting mix moist but not drenched.

In the cooler months, water less; allow the potting mix to dry out for one to two weeks between watering.

Back-deck planting

Adding clusters of plants to a back deck is a great way to frame the view of the yard beyond. I also like using plants to feather the relationship between indoors and outdoors. It allows our interiors to link naturally to the outside space. There are certain plants I sway towards for seasonality and atmosphere. This recipe explores autumn tones and rustic textures.

 Partial shade to partial sun

KIT OF PARTS

PLANTS

1 Japanese maple (*Acer palmatum*)

7 Sikkim creepers (*Parthenocissus sikkimensis*)

2 paperplants (*Fatsia japonica*)

5 leatherleaf sedges (*Carex buchananii*)

PLANTERS

1 extra-large grey planter, approx. 50 cm (20 in) diameter and 50 cm (20 in) high

2 large grey planters, approx. 40 cm (16 in) diameter and 40 cm (16 in) high

1 medium grey planter, approx. 30 cm (12 in) diameter and 30 cm (12 in) high

1 small grey planter, approx. 20 cm (8 in) diameter and 20 cm (8 in) high

PLANTING

premium-quality general-purpose potting mix

METHOD

Position the extra-large planter in one corner of your deck. Place a layer of potting mix into the planter, approximately 10–20 centimetres (4–8 inches) deep, and place the Japanese maple.

Backfill with potting mix, leaving enough room to plant three Sikkim creepers evenly around the Japanese maple. Continue to backfill with potting mix, making sure to gently press down to compact the potting mix.

Place a large planter to one side of the extra-large planter. Plant one paperplant and two Sikkim creepers into this planter. Place the small planter in front of these two planters so that it overlaps the two. Plant two leatherleaf sedges into the small planter.

Position the second large planter in the other corner of the deck, or simply apart from the first cluster of planters. Plant one paperplant and two Sikkim creepers into this planter. Then place the medium planter slightly offset from this planter. Plant three leatherleaf sedges into the medium planter.

Finish by watering in your plants.

WATERING

In the warmer months, keep the potting mix moist but not drenched.

In the cooler months, water less; allow the potting mix to dry out for approximately one week between watering.

Fresh TV console

As much as we try to hide the TV, it's a difficult task to achieve. Instead, I like to soften the edges around the console that the TV sits on or hangs above.

 Filtered to bright light

KIT OF PARTS

PLANTS

1 blue star fern (*Phlebodium aureum*)

1 wax plant (*Hoya* sp.)

1 maidenhair fern (*Adiantum* sp.)

PLANTERS

1 small off-white planter, approx. 20 cm (8 in) diameter and 20 cm (8 in) high

2 extra-small off-white planters, approx. 15 cm (6 in) diameter and 15 cm (6 in) high

PLANTING

premium-quality indoor potting mix

PREPARATION

Ensure that your planters have drip trays to prevent your console from becoming stained.

METHOD

Place a layer of potting mix into the small planter, approximately 5–10 centimetres (2–4 inches) deep, and place the blue star fern. Backfill with potting mix, making sure to gently press down to compact the potting mix.

Next, plant the wax plant and maidenhair fern into the two extra-small planters.

To style, position the small planter with the blue star fern to one side of your TV. Place the maidenhair fern between the blue star fern and the TV to provide a soft buffer, then place the wax plant on the other side of the blue star fern.

Finish by watering in your plants.

WATERING

In the warmer months, keep the potting mix moist but not drenched.

In the cooler months, water less; allow the potting mix to dry out for approximately one week between watering.

STYLING TIP

FEEL FREE TO USE COLOURED PLANTERS, BUT ENSURE THAT THEY MATCH AND THEY DON'T HAVE DISTRACTING PATTERNS.

A love of burgundy

Not all gardens need to be green. Here, I've experimented with burgundy tones and a splash of grey in a courtyard to see if these colours would work. (They do!) You can play with colour combinations in your space by trialling small patches or moving your planters around to see how the combinations look.

 Partial sun to full sun

KIT OF PARTS

PLANTS

1 *Cotinus coggygria* 'Grace'

3 tree houseleeks (*Aeonium* spp.)

2 *Ligularia dentata* 'Pandora'

2 *Agastache* 'Sweet Lili'

2 lavender cottons (*Santolina* spp.)

PLANTERS

1 extra-large charcoal planter, approx. 50 cm (20 in) diameter and 50 cm (20 in) high

2 large charcoal planters, approx. 40 cm (16 in) diameter and 40 cm (16 in) high

1 medium charcoal planter, approx. 30 cm (12 in) diameter and 30 cm (12 in) high

1 small charcoal planter, approx. 20 cm (8 in) diameter and 20 cm (8 in) high

PLANTING

premium-quality general-purpose potting mix

METHOD

Position the extra-large planter in the back corner of your courtyard. Place a layer of potting mix into the planter, approximately 20–30 centimetres (8–12 inches) deep, and place the *Cotinus coggygria* 'Grace' plant in the centre. Backfill with potting mix, making sure to gently press down to compact the potting mix.

Position one of the large planters next to the extra-large planter, and plant one tree houseleek, one *Ligularia dentata* 'Pandora' and one *Agastache* 'Sweet Lili'.

Place the second large planter on the other side of the courtyard and repeat the plantings. Backfill both large planters with potting mix, making sure to gently press down to compact the potting mix.

Place the medium planter to the left of the lone large planter, and pot up one lavender cotton.

Place the small planter in front of the extra-large planter. Plant one tree houseleek and the remaining lavender cotton. Backfill both the medium and small planters with potting mix, making sure to gently press down to compact the potting mix.

Finish by watering in your plants.

WATERING

In the warmer months, keep the potting mix moist but not drenched. During summer, you may need to water once or twice a day for your plants to thrive.

In the cooler months, water less; allow the potting mix to dry out for a couple of days between watering.

Potted seasonality

We all remember our grandparents creating potted moments of seasonal flowering plants. I must admit that I have a soft spot for them. They make an ideal addition if you're after something that can be changed throughout the seasons, and they can be tailored to the colours and flowers you love the most.

 Partial sun to full sun

KIT OF PARTS

PLANTS

1 avens (*Geum* sp.)

1 orange New Zealand sedge (*Carex testacea*)

2 *Erigeron* 'LA Pink'

3 tulip bulbs – red-, pink- or orange-flowering tulips

6 kidney weeds (*Dichondra repens*)

PLANTER

1 low, wide planter approx. 40 cm (16 in) diameter and 10 cm (4 in) high

PLANTING

premium-quality general-purpose potting mix

slow-release fertiliser

METHOD

Place a layer of potting mix into the planter, approximately 2–5 centimetres (¾–2 inches) deep. Before placing your plants in the planter, sprinkle in a layer of slow-release fertiliser. Add another layer of potting mix.

Position the avens to one side of the planter. Place the orange New Zealand sedge next to it, towards the centre of the planter. Place one *Erigeron* 'LA Pink' plant on either side of the orange New Zealand sedge and avens.

Plant the three tulip bulbs on the other side of the planter. Backfill with potting mix, leaving enough room to plant the kidney weeds in any gaps in the planter. Continue to backfill the planter, making sure to gently press down to compact the potting mix.

Finish by watering in your plants.

WATERING

In the warmer months, keep the potting mix moist but not drenched. During summer, you may need to water once or twice a day for your plants to thrive.

In the cooler months, water less; allow the potting mix to dry out for a couple of days between watering.

Bowls of bulbs

Explosions of colour can be added effectively through easy-to-create bowls and planters of the same bulb. My favourites are tulips, dahlias and daffodils (*Narcissus* spp.). I use these around the garden and in the house to alter the atmosphere throughout the seasons. It always feels like a little surprise when they start flowering.

 Partial sun to full sun

KIT OF PARTS

PLANTS

30–40 daffodils (*Narcissus* spp.) in a variety of colours, or a similar number of bulbs of your choice (if you're using bigger bulbs, then you may need fewer; for instance, you'll require fewer dahlia bulbs [tubers] to fill the same planters)

PLANTERS

1 low terracotta planter, approx. 30 cm (12 in) diameter and 15 cm (6 in) high

1 standard ceramic planter, approx. 40 cm (16 in) diameter and 20 cm (8 in) high

PLANTING

premium-quality general-purpose potting mix

slow-release fertiliser

METHOD

Fill both planters with potting mix, stopping approximately 5 centimetres (2 inches) from the top of the planters. Sprinkle in a layer of slow-release fertiliser, and add another layer of potting mix.

Position all of your bulbs so they're evenly spaced. Once you're happy with the spacing, backfill the planters with potting mix, making sure to gently press down to compact the potting mix.

Style these planters by placing them with other clusters in your garden or on outdoor or indoor tables.

Finish by watering in your plants.

WATERING

In the warmer months, keep the potting mix moist but not drenched. During summer, you may need to water once or twice a day for your plants to thrive.

In the cooler months, place the planters where they can receive rainfall while the bulbs are dormant. I like to place mine under a tree.

YOU CAN PLANT YOUR BULBS LOOSELY OR CLOSE TOGETHER, DEPENDING ON HOW TIGHT YOU WANT THE FLOWER CLUSTERS TO BE.

CHAPTER SIX

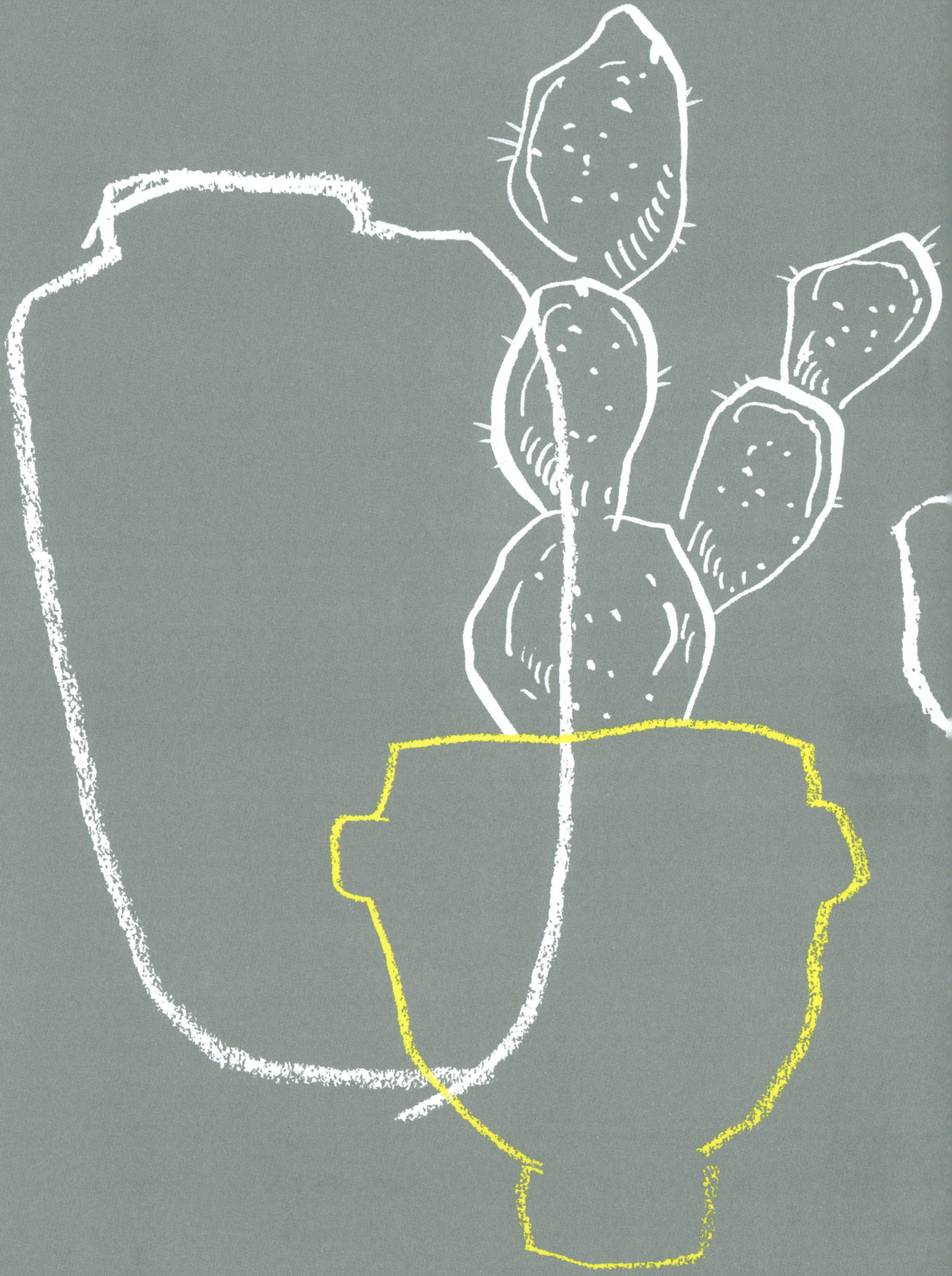

blueprints

The projects in this chapter will help you conceptualise and create your own garden, tying in all the elements of what you've learned so far.

HOW TO READ
A BLUEPRINT

These garden blueprints provide step-by-step instructions to help you create your own designed garden. The plan sketches will guide you with plant placement; they take into consideration not only things such as available light and growing habits, but also the overall design aesthetic of the space and the relationship of the plants to the built structures and other features.

As with the recipes, if you fall in love with an idea but want to use it in a space that's different from the one pictured, then feel free to adapt it, using the information in this book to guide you.

OPTIMAL LIGHTING CONDITIONS

A guide to the lighting range that best suits the plants in the garden project.

TIP

This extra information will help with styling and looking after your plants.

CULTIVATION

Pruning and other maintenance advice for the plants used in the garden project.

THE PLANT SOCIETY DESIGN HANDBOOK

The perfect greenhouse

Many gardeners dream of having their own greenhouse. These warm, sheltered structures create the perfect microclimate to grow exotic species that wouldn't thrive outdoors. However, home greenhouses don't need to be used purely to house exotic specimens; they can allow you to plant produce earlier in the season so that your edible plants have had a head start before the weather picks up. I love mixing special tropical plants with my productive garden seedlings.

Partial shade to full sun

KIT OF PARTS

PLANTS

3 orchids – choose your own preferred species

2 orchid cacti (*Epiphyllum* spp.)

1 tassel fern (*Huperzia* sp.)

2 mistletoe cacti (*Rhipsalis* spp.)

assorted produce or flower seeds

PLANTERS

3 medium terracotta planters, approx. 30 cm (12 in) diameter and 30 cm (12 in) high

2 medium hanging planters, approx. 20 cm (8 in) diameter and 20 cm (8 in) high

3 small hanging planters, approx. 14 cm (5½ in) diameter and 14 cm (5½ in) high

seed-raising trays or punnets as required

PLANTING

orchid bark

premium-quality potting mix

seed-raising mix

METHOD

Place a layer of orchid bark in a medium terracotta planter, approximately 5–15 centimetres (2–6 inches) deep. Place an orchid into the planter, and backfill with orchid bark. You can gently compact the bark, but it is important to have air pockets in the mix for the orchid roots to grow into.

Repeat for the remaining medium terracotta planters and orchids. Place the potted orchids on your growing bench so they're not on the cold floor.

Place a layer of potting mix into a medium hanging planter, approximately 5–10 centimetres (2–4 inches) deep, and place an orchid cactus. Backfill with potting mix, making sure to gently press down to compact the potting mix. Repeat for the other medium hanging planter and orchid cactus.

Place a layer of orchid bark into one of the small hanging planters, approximately 5–8 centimetres (2–3 inches) deep, and place the tassel fern. Backfill with more orchid bark.

Place a layer of potting mix into the remaining small hanging planters, approximately 5–8 centimetres (2–3 inches) deep, and place a mistletoe cactus into each one. Backfill with potting mix, making sure to gently press down to compact the potting mix.

Hang all of your hanging planters from the ceiling of your greenhouse so that the medium and small planters are layered in an organic way. Plant your produce or flower seeds into your seed-raising trays or punnets. Refer to the planting guide on your seed packets.

Finish by watering in your potted plants.

WATERING

In the warmer months, water once or twice a day. If a heatwave is expected, then increase watering as required.

In the cooler months, water as required.

232

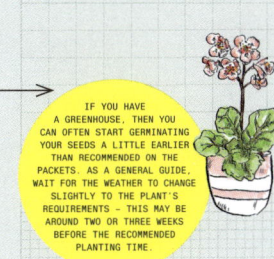

IF YOU HAVE A GREENHOUSE, THEN YOU CAN OFTEN START GERMINATING YOUR SEEDS A LITTLE EARLIER THAN RECOMMENDED ON THE PACKETS. AS A GENERAL GUIDE, WAIT FOR THE WEATHER TO CHANGE SLIGHTLY TO THE PLANT'S REQUIREMENTS – THIS MAY BE AROUND TWO OR THREE WEEKS BEFORE THE RECOMMENDED PLANTING TIME.

CULTIVATION

Orchids: Remove spent flowers and dead foliage as needed.

Orchid cactus: Trim dead ends of foliage and remove spent flowers as they occur.

Tassel fern, mistletoe cactus: Remove dead leaves by using your fingers or pruning them with small snips.

Seeds: As they're tender during the germination stage, make sure to keep the soil moist but not drenched at all times.

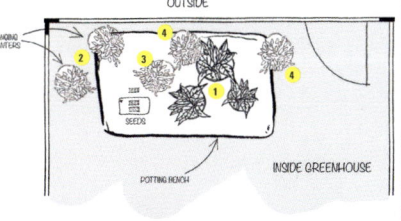

01 orchid – your own preferred species

02 orchid cactus (*Epiphyllum* sp.)

03 tassel fern (*Huperzia* sp.)

04 mistletoe cactus (*Rhipsalis* sp.)

233

KIT OF PARTS

Everything you need for your garden project, including plants, planters, soil type, compost, mulch and gravel.

METHOD

A step-by-step project guide, including preparation, planting and watering tips. (For advice on planting, see pages 112–15.)

KEY

The site plan legend that shows the plants used in the garden project.

BLUEPRINT

The sketched plan maps out an aerial view of the plants in the space.

Suburban escape

The thought of tending a garden can be daunting, but with the right plant selection you can cut down on the maintenance required to keep your garden looking great. This combination of plants works especially well in small courtyards and compact front yards, and the whimsical, less-manicured look will create a soothing oasis.

 Partial shade to full sun

KIT OF PARTS

PLANTS

3 giant hyssops (*Agastache* spp.)

3 *Calamagrostis* × *acutiflora* 'Karl Foerster'

8 lamb's ears (*Stachys byzantina*)

8 kidney weeds (*Dichondra repens*)

5 native violets (*Viola hederacea*)

5 elephant's ears (*Bergenia* spp.)

3 catnips (*Nepeta cataria*)

3 tractor seat plants (*Cremanthodium reniforme*)

3 rosemaries (*Salvia rosmarinus*)

PLANTING

compost

premium-quality garden soil

stepping stones

crushed rock or road base

mulch

PREPARATION

These plants will cover a 5-metre (16-foot)-long by 4-metre (13-foot)-wide garden bed. If your garden bed is larger, then simply repeat the pattern with more plants.

Prepare your garden bed by turning the existing soil, then adding compost and additional premium-quality garden soil (see page 111).

METHOD

Tentatively space out the stepping stones. When you're happy with the layout of the stepping stones, lift them up one by one. Remove enough soil to lay a bed of crushed rock or road base to support the stepping stones. A thickness of approximately 2 centimetres ($^3/_4$ inch) is ideal. Compact the crushed rock or road base into a flat bed, and lay the stepping stones directly on top.

Lay out the taller plants (the giant hyssops and *Calamagrostis* × *acutiflora* 'Karl Foerster' plants). This will allow you to visualise the highs and lows in the garden.

Place the remaining plants as per the blueprint to the right. When you're happy with the form of the layout, you can plant your garden bed.

Add a layer of mulch approximately 5–10 centimetres (2–4 inches) thick.

Finish by watering in your plants.

WATERING

In the warmer months, water once a week. If a heatwave is expected, then increase watering as required.

In the cooler months, water as required.

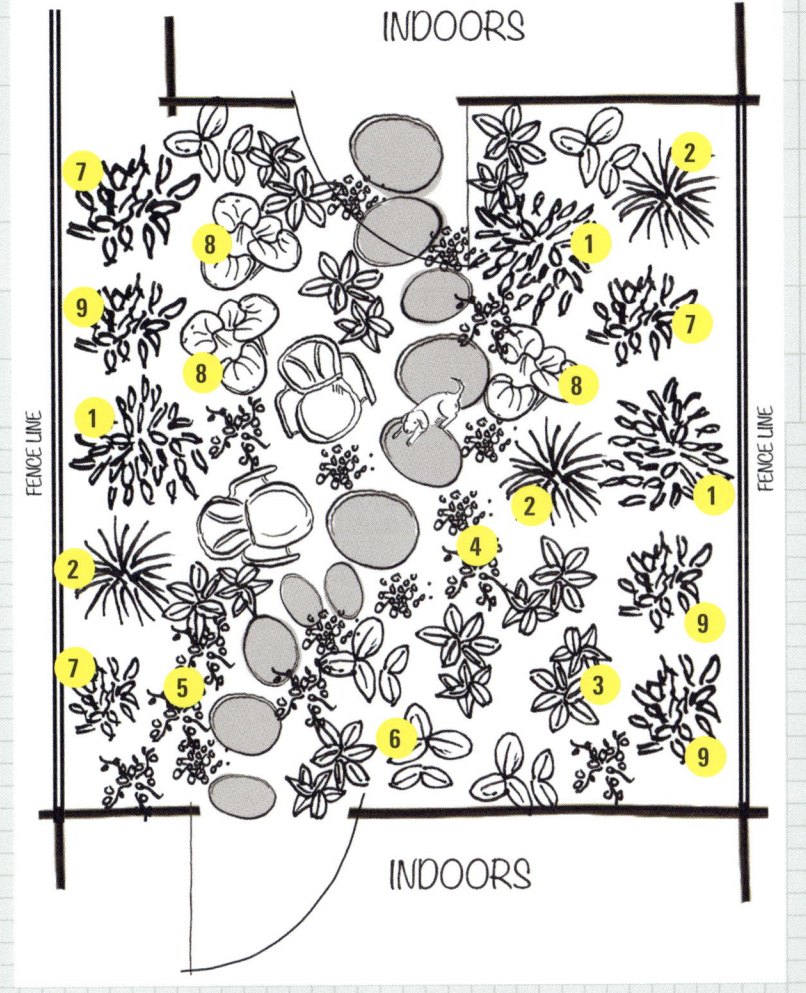

INDOORS

FENCE LINE

FENCE LINE

INDOORS

CULTIVATION

Giant hyssop, *Calamagrostis × acutiflora* **'Karl Foerster':** Cut back after the last frost.
Lamb's ear, catnip: Deadhead the flowers once they're spent.
Kidney weed, native violet: Trim back as desired to control growth.
Elephant's ear, tractor seat plant: Remove old foliage as the plants die back.
Rosemary: Shape as required.

01
giant hyssop
(*Agastache* sp.)

02
Calamagrostis × acutiflora 'Karl Foerster'

03
lamb's ear
(*Stachys byzantina*)

04
kidney weed
(*Dichondra repens*)

05
native violet
(*Viola hederacea*)

06
elephant's ear
(*Bergenia* sp.)

07
catnip
(*Nepeta cataria*)

08
tractor seat plant
(*Cremanthodium reniforme*)

09
rosemary
(*Salvia rosmarinus*)

Green link between areas

Open-plan living provides the ideal canvas when curating an indoor potted garden. A cluster of plants can act as a screen, or help delineate areas within a larger space. For larger indoor gardens that are more than a couple of clusters, I suggest choosing a few plant types and repeating them en masse.

 Filtered to bright light

KIT OF PARTS

PLANTS

2 kauri pines (*Agathis robusta*)

3 broadleaf lady palms (*Rhapis excelsa*)

13 *Cissus* 'Ellen Danica'

PLANTERS

2 extra-large charcoal planters, approx. 50 cm (20 in) diameter and 50 cm (20 in) high

3 large charcoal planters, approx. 40 cm (16 in) diameter and 40 cm (16 in) high

PLANTING

premium-quality potting mix

PREPARATION

Ensure that your planters have drip trays to prevent your floor from becoming stained.

METHOD

Lay out the planters. Use the blueprint to the right as a guide. You can easily alter the layout to suit your living and dining space.

When you're happy with the arrangement, plant the kauri pines into the two extra-large planters, and plant the broadleaf lady palms into the three large planters.

Backfill with potting mix, leaving enough room to plant two or three *Cissus* 'Ellen Danica' plants around the edge of each planter. Continue to backfill with potting mix, making sure to gently press down to compact the potting mix.

Finish by watering in your plants.

WATERING

In the warmer months, water once a week.

In the cooler months, water every two weeks.

CULTIVATION

Kauri pine, *Cissus* 'Ellen Danica': Remove any dead leaves by hand.
Broadleaf lady palm: Trim back dead foliage as it appears.

LIVING ROOM

DINING ROOM

WINDOW

WINDOW

01
kauri pine
(*Agathis robusta*)

02
broadleaf lady palm
(*Rhapis excelsa*)

03
Cissus 'Ellen Danica'

Green-wall shelves

With space in our cities becoming scarcer, we're experimenting with creative ways to incorporate gardens into our living spaces. Sprawling across exterior facades and adorning internal walls, green walls not only enhance aesthetics but also allow for nature to be included in our manufactured spaces. However, a good green wall requires detailed planning, care and construction, making it unachievable for many people. An alternative way to use plants on the vertical plane is to employ a shelving system. It offers easy maintenance while providing versatility by allowing for a changing layout.

 Filtered to bright light

KIT OF PARTS

PLANTS

5 mistletoe cacti (*Rhipsalis* spp.)

5 heartleaf philodendrons (*Philodendron hederaceum*)

3 *Cissus* 'Ellen Danica'

3 wax plants (*Hoya* spp.)

3 mini monsteras (*Rhaphidophora tetrasperma*)

3 dragon-tail plants (*Epipremnum pinnatum*)

3 arrowhead plants (*Syngonium* spp.)

PLANTERS

10 medium self-watering planters, approx. 25 cm (10 in) diameter and 25 cm (10 in) high

15 small self-watering planters, approx. 20 cm (8 in) diameter and 20 cm (8 in) high

PLANTING

premium-quality indoor potting mix

METHOD

Repot your plants into the self-watering planters. Place one mistletoe cactus or heartleaf philodendron into each medium planter, and one of the other plants into each of the small planters.

Plan your pattern so that your green-wall shelves feel considered rather than random. Start by placing plants on the bottom shelf and then move upwards. On the bottom row, I recommend placing one or two heartleaf philodendrons – this plant cascades downwards as it grows and can cope with lower light levels.

Try to cluster the plant types together so that your green-wall shelves don't feel chaotic. For instance, I recommend placing two of the same type on the same shelf and then placing one of the same plant on the shelf above or below. Plants that suit the top of your shelving system are mini monsteras, dragon-tail plants and arrowhead plants. These tend to grow upwards as well as downwards.

Finish by watering in your plants.

WATERING

In the warmer months, water once a week.

In the cooler months, water every two weeks.

CULTIVATION

Areas of your green-wall shelves will receive different levels of natural light. The benefit of using shelving as a green wall is having the flexibility to rotate the planters regularly. As your cascading plants grow, you may need to prune them back so they don't swallow up the plants below them. This allows the natural light to continue to reach each plant on your shelves.

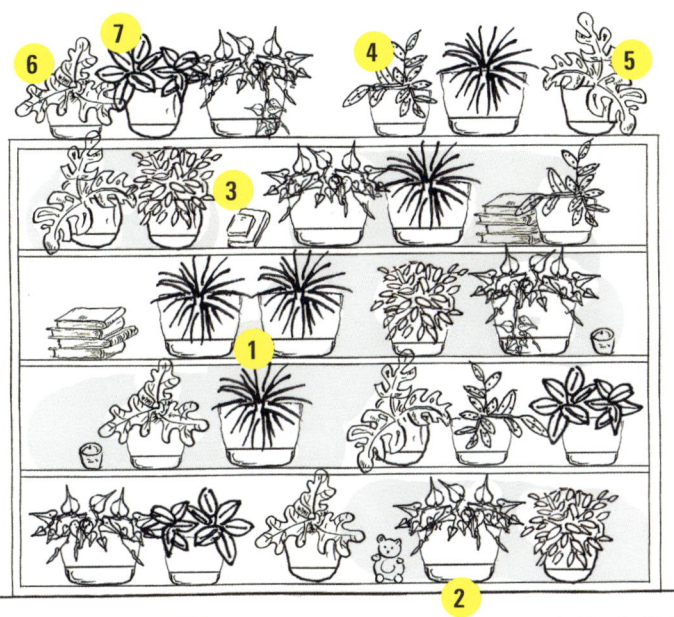

CEILING

You can play with the placement of plants, books, vases and other shelf items until you achieve your desired look.

THE SELF-WATERING PLANTERS CAN BE ANY COLOUR YOU LIKE, BUT THEY'LL CREATE VISUAL UNITY ACROSS THE GREEN-WALL SHELVES IF THEY'RE ALL THE SAME TONE.

01
mistletoe cactus (*Rhipsalis* sp.)

02
heartleaf philodendron (*Philodendron hederaceum*)

03
Cissus 'Ellen Danica'

04
wax plant (*Hoya* sp.)

05
mini monstera (*Rhaphidophora tetrasperma*)

06
dragon-tail plant (*Epipremnum pinnatum*)

07
arrowhead plant (*Syngonium* sp.)

Indoor winter garden

 Filtered to bright light

KIT OF PARTS

PLANTS

1 banana plant (*Musa* sp.)

1 tailflower (*Anthurium* sp.)

2 elephant's ears (*Alocasia* spp.)

3 spotted begonias (*Begonia maculata*)

1 flat-leaved vanilla (*Vanilla planifolia*)

assorted vegetable seeds for germinating in the cooler months

PLANTERS

1 extra-large charcoal planter, approx. 50 cm (20 in) diameter and 90 cm (35 in) high

2 medium charcoal planters, approx. 30 cm (12 in) diameter and 30 cm (12 in) high

2 small charcoal planters, approx. 20 cm (8 in) diameter and 20 cm (8 in) high

3 extra-small charcoal planters, approx. 15 cm (6 in) diameter and 15 cm (6 in) high

2 seed-germination kits

2 rattan or cane baskets

PLANTING

premium-quality indoor potting mix

seed-raising mix

vermiculite

Indoor winter gardens are a gardener's dream. Much like a greenhouse, they allow you to garden while sheltered from the weather. They also extend the living space into a quasi-outdoor room, allowing you to bring the garden inside. Another advantage of these special rooms is that they protect fragile plants from cold weather, so they're an ideal space to grow plants that prefer warmer and tropical conditions.

PREPARATION

Ensure that your planters have drip trays to prevent your floor from becoming stained.

METHOD

Place a layer of potting mix into the extra-large planter, approximately 40–50 centimetres (16–20 inches) deep, and place the banana plant. Backfill with potting mix, making sure to gently press down to compact the potting mix.

Plant the tailflower into one of the medium planters, and plant one elephant's ear into the other medium planter.

Plant the other elephant's ear into one of the small planters, and plant one spotted begonia into the other small planter. Once potted up, these small planters will sit in the two rattan or cane baskets. The two remaining spotted begonias and the flat-leaved vanilla are each planted into a separate extra-small planter.

Plant your chosen vegetable seeds into germination trays using seed-raising mix and vermiculite (see page 130). I like to grow seeds inside at the end of winter to give my seedlings a head start. Once planted, place the covers on to help insulate the seeds (and soon-to-be seedlings).

Arrange the planters and seed trays as per the blueprint to the right.

Finish by watering in your plants.

WATERING

In the warmer months, water once a week.

In the cooler months, water every two weeks.

CULTIVATION

Banana plant: It's often dormant in winter. Simply cut back any dead foliage, and keep it away from frost during the cooler months.

Tailflower, flat-leaved vanilla: Remove spent foliage and flowers as needed.

Elephant's ear: During the cooler months, it can die back to the corm. Avoid drenching the soil, otherwise the corm may rot.

Spotted begonia: Maintain good airflow to prevent powdery mildew from developing on the foliage.

Vegetable seeds: When your seedlings are ready, plant them out in the garden.

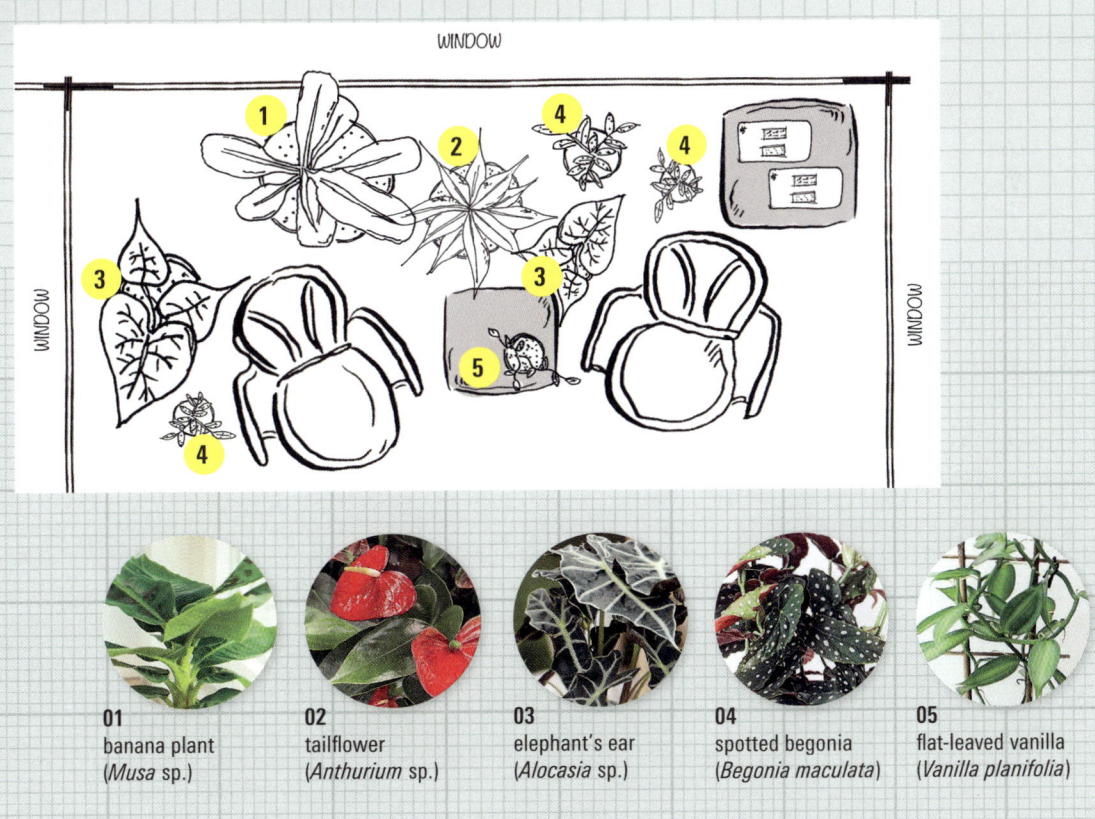

01 banana plant (*Musa* sp.)

02 tailflower (*Anthurium* sp.)

03 elephant's ear (*Alocasia* sp.)

04 spotted begonia (*Begonia maculata*)

05 flat-leaved vanilla (*Vanilla planifolia*)

Lush narrow fence line

Whether you live in a suburban home or an urban terrace house, narrow strips of soil beside your home are often disused. These impractical spaces can easily be turned into moments of greenery that are a joy to behold, while also creating an environment that attracts birds. Typically, being shaded spaces, they require plant species that belong in such ecosystems. When imagining a garden with these qualities, I draw inspiration from rainforest floors and woodland gardens.

 Partial shade to partial sun

KIT OF PARTS

PLANTS

3 silver vein creepers (*Parthenocissus henryana*)

5 winter roses (*Helleborus orientalis*)

5 cast iron plants (*Aspidistra elatior*)

15 kidney weeds (*Dichondra repens*)

15 native violets (*Viola hederacea*)

PLANTING

compost

premium-quality garden soil

mulch

PREPARATION

These plants will cover a 3-metre (10-foot)-long by 1-metre (3-foot)-wide garden bed. If your garden bed is larger, then simply repeat the pattern with more plants.

Prepare your garden bed by turning the existing soil, then adding compost and additional premium-quality garden soil (see page 111).

METHOD

Tentatively place the silver vein creepers along the fence line or wall. Lay out the winter roses and cast iron plants so you can see their form. Place the remaining plants as per the blueprint to the right. When you're happy with the form of the layout, you can plant your garden bed. Add a layer of mulch approximately 5–10 centimetres (2–4 inches) thick.

Finish by watering in your plants.

WATERING

In the warmer months, water once a week.

In the cooler months, water every two weeks.

CULTIVATION

Silver vein creeper: Prune back to manage growth. (For pruning tips, see pages 124–9.)
Winter rose: As the foliage dies, trim it back to keep the plant clean.
Cast iron plant: Trim away dead foliage as it appears.
Kidney weed, native violet: Trim back as desired to control growth.

Continue this pattern of planting along the fence line for maximum impact.

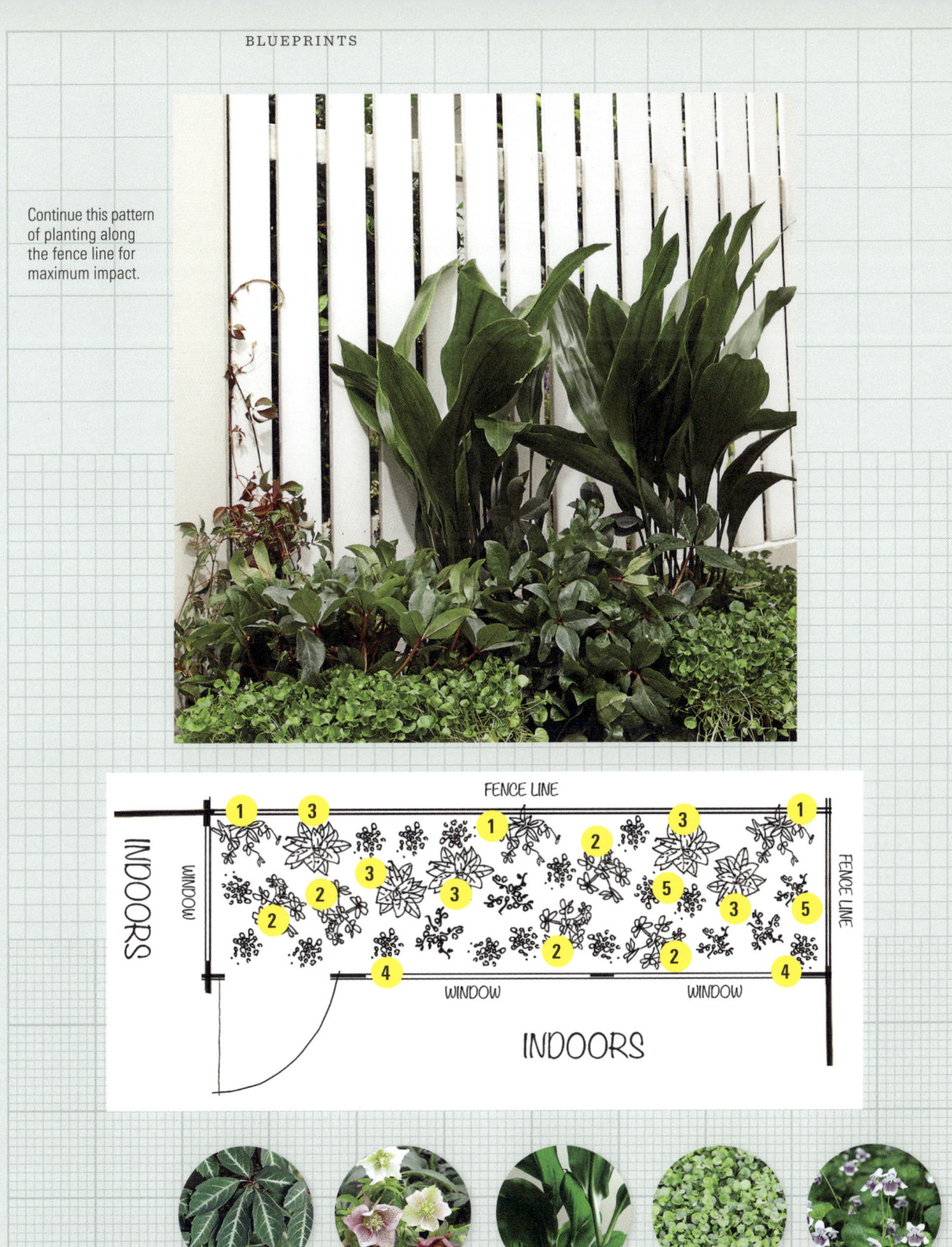

01
silver vein creeper
(*Parthenocissus henryana*)

02
winter rose
(*Helleborus orientalis*)

03
cast iron plant
(*Aspidistra elatior*)

04
kidney weed
(*Dichondra repens*)

05
native violet
(*Viola hederacea*)

Sheltered space

Shady areas between buildings can be uninspiring when curating a garden. The lack of full sun makes them challenging when workshopping what plants to establish. If you have a courtyard that doesn't get much sun, then take a moment to find a similar scenario in the natural world (such as woodlands and rainforests). The plants in these environments have evolved to thrive in sheltered positions, where light reaches the foliage through a canopy of trees and ferns, and blooms such as plantain lilies (*Hosta* spp.) and Japanese windflowers (*Eriocapitella hupehensis*) sway gracefully in the breeze.

 Partial shade to partial sun

KIT OF PARTS

PLANTS

1 Japanese maple (*Acer palmatum*)

10 Japanese windflowers (*Eriocapitella hupehensis*)

7 Japanese forest grasses (*Hakonechloa macra*)

7 plantain lilies (*Hosta* spp.)

20 kidney weeds (*Dichondra repens*)

PLANTING

compost

premium-quality garden soil

coarse gravel for drainage

mulch

PREPARATION

These plants will cover a 3-metre (10-foot)-long by 3-metre (10-foot)-wide garden bed. If your garden bed is larger, then simply repeat the pattern with more plants.

Prepare your garden bed by turning the existing soil, then adding compost, additional premium-quality garden soil and coarse gravel for drainage (see page 111).

METHOD

Place the Japanese maple in the centre of the courtyard. This will become the hero of your garden. Space out the Japanese windflowers and Japanese forest grasses around the Japanese maple. Space out the plantain lilies among the other plants. You're aiming for a good mix of chunky versus fine foliage. Space out the kidney weeds. Don't be too worried about the exact position of these plants, as they'll fill the floor of the courtyard over time.

When you're happy with the form of the layout, you can plant your courtyard. Add a layer of mulch approximately 5 centimetres (2 inches) thick.

Finish by lightly watering in your plants.

WATERING

In the warmer months, water once a fortnight to once a month.

In the cooler months, water as required.

CULTIVATION

Japanese maple: As the leaves drop, rake them up and place them in the compost.
Japanese windflower, Japanese forest grass, plantain lily: Cut back the dead foliage as the plants die back.
Kidney weed: Trim back as desired to control growth.

FENCE LINE

FENCE LINE

FENCE LINE

③

②

①

④

⑤

INDOORS

You can modify the design by placing the Japanese maple next to the house, rather than in the centre of the courtyard.

01
Japanese maple
(*Acer palmatum*)

02
Japanese windflower
(*Eriocapitella hupehensis*)

03
Japanese forest grass
(*Hakonechloa macra*)

04
plantain lily
(*Hosta* sp.)

05
kidney weed
(*Dichondra repens*)

Small edible garden

Gardens benefit from the addition of edibles, even if they're simply for aromatic purposes. I use creeping thyme and creeping rosemary in my designs to form lush, fragrant edges in a range of spaces. Herbs are aesthetically versatile in the garden while also being useful in the kitchen.

 Partial sun to full sun

KIT OF PARTS

PLANTS

1 lemon tree (*Citrus limon*)

7 creeping rosemaries (*Rosmarinus officinalis* Prostratus Group)

1 cumquat tree (*Citrus japonica*)

1 foxtail agave (*Agave attenuata*)

1 twin-flowered agave (*Agave geminiflora*)

6 creeping thymes (*Thymus serpyllum*)

PLANTERS

1 extra-large terracotta planter, approx. 50 cm (20 in) diameter and 50 cm (20 in) high

1 large terracotta planter, approx. 40 cm (16 in) diameter and 40 cm (16 in) high

2 medium terracotta planters, approx. 30 cm (12 in) diameter and 30 cm (12 in) high

1 small terracotta planter, approx. 20 cm (8 in) diameter and 20 cm (8 in) high

PLANTING

premium-quality potting mix

mulch

METHOD

Place the extra-large planter in a corner of your space. I like using corners that are underutilised to add some life.

Place a layer of potting mix into the extra-large planter, approximately 30–40 centimetres (12–16 inches) deep, and place the lemon tree. Backfill with potting mix, leaving enough room to plant three creeping rosemaries. Add the creeping rosemaries, then continue to backfill with potting mix, making sure to gently press down to compact the potting mix.

Position the large planter beside the extra-large planter. Place a layer of potting mix into the planter, approximately 20–30 centimetres (8–12 inches) deep, and place the cumquat tree. Backfill with potting mix, and underplant with three creeping rosemaries.

Place the two medium planters either side of the cluster. Add a small layer of potting mix, and plant the foxtail agave in one and the twin-flowered agave in the other. Backfill both planters with potting mix, leaving enough room to plant three creeping thymes in each one. Add the creeping thymes, then continue to backfill with potting mix, making sure to gently press down to compact the potting mix.

Place the small planter to one side of a medium planter. Place a thin layer of potting mix in the bottom, and plant a creeping rosemary into the planter. Backfill with potting mix, and gently press down to compact the potting mix.

Finish by watering in your plants.

WATERING

In the warmer months, water once or twice a day.

In the cooler months, water as required.

CULTIVATION

Lemon tree, cumquat tree: Citrus trees are big feeders, so fertilise with a good-quality citrus food as per the instructions on the packaging.

Creeping rosemary, creeping thyme: Tip prune to promote denser growth.

Foxtail agave, twin-flowered agave: Trim dead leaves and flowers as they appear.

STEPS

6 4
1 3 2
5

TO HELP PRESERVE THE MOISTURE IN THE POTTING MIX WHEN USING TERRACOTTA PLANTERS, YOU CAN SEAL YOUR PLANTERS WITH A POT SEALER.

01
lemon tree
(*Citrus limon*)

02
creeping rosemary
(*Rosmarinus officinalis*
Prostratus Group)

03
cumquat tree
(*Citrus japonica*)

04
foxtail agave
(*Agave attenuata*)

05
twin-flowered agave
(*Agave geminiflora*)

06
creeping thyme
(*Thymus serpyllum*)

227

Dappled garden

I like to find a way to grow plants even in the most difficult of situations. Light wells can be a challenge when gardening because of their lack of bright light. However, if we look at the natural world, we can quickly find growing scenarios that are very similar to our built environment. You can easily turn what may appear to be a hopeless space into one that mimics the rainforest floor.

 Full shade to partial sun

KIT OF PARTS

PLANTS

5 tractor seat plants
(*Cremanthodium reniforme*)

3 rabbit's foot ferns (*Davallia solida* var. *fejeensis*)

7 kidney weeds (*Dichondra repens*)

5 native violets (*Viola hederacea*)

PLANTING

compost

premium-quality garden soil

mulch

PREPARATION

These plants will cover a 1-metre (3-foot)-long by 1-metre (3-foot)-wide garden bed. If your garden bed is larger, then simply repeat the pattern with more plants.

Prepare your garden bed by turning the existing soil, then adding compost and additional premium-quality garden soil (see page 111).

METHOD

Tentatively lay out the tractor seat plants as per the blueprint to the right.

Lay out the rabbit's foot ferns, kidney weeds and native violets as per the blueprint to the right.

When you're happy with the form of the layout, you can plant your garden bed.

Add a layer of mulch approximately 5–10 centimetres (2–4 inches) thick.

Finish by watering in your plants.

WATERING

In the warmer months, water once or twice a week. If a heatwave is expected, then increase watering as required.

In the cooler months, water as required.

CULTIVATION

Tractor seat plant, rabbit's foot fern: Remove dead foliage as required.
Kidney weed, native violet: Trim back as desired to control growth.

In this small section of
the garden, the huge
leaves of the tractor seat
plants contrast with
the delicate foliage of
the kidney weeds and
rabbit's foot ferns.

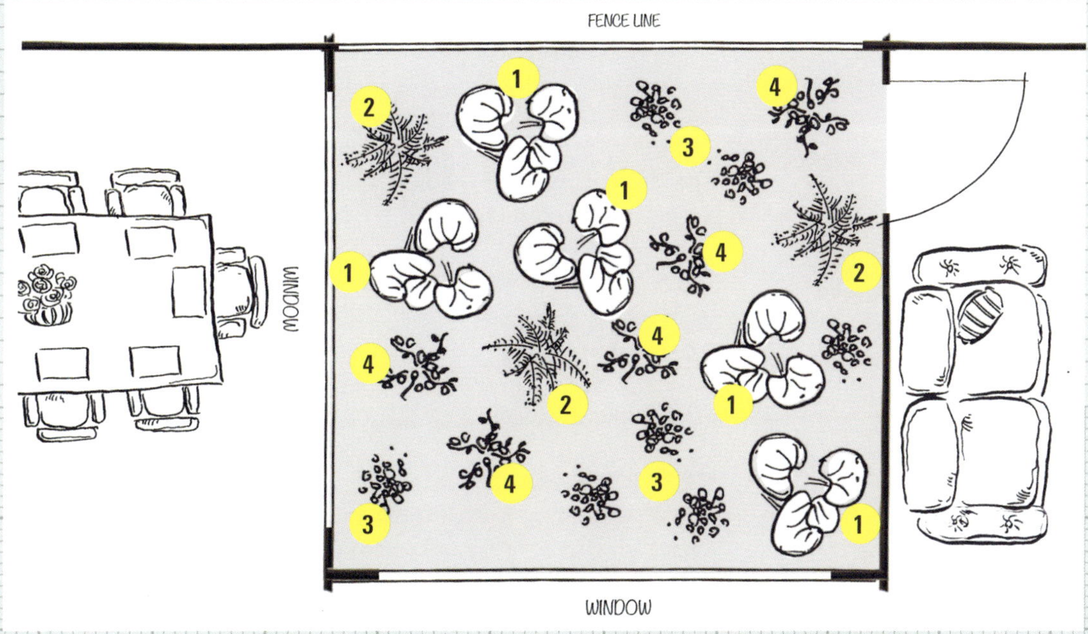

FENCE LINE

WINDOW

WINDOW

01 tractor seat plant
(*Cremanthodium
reniforme*)

02 rabbit's foot fern
(*Davallia solida* var.
fejeensis)

03 kidney weed
(*Dichondra repens*)

04 native violet
(*Viola hederacea*)

Fuss-free front yard

Front yards provide a clue to the home's inhabitants. They're a welcoming gesture and don't have to be high maintenance or confronting when it comes to care. If you're curating a slightly bigger garden, then try to embrace a more naturalistic approach to your aesthetic and allow your plants to look less manicured. Gardens need not be prim and proper – there is beauty in your plants going through the seasonal shifts.

 Full shade to partial sun

KIT OF PARTS

PLANTS

5 Chinese silver grasses (*Miscanthus sinensis*)

3 honey bushes (*Melianthus major*)

7 giant hyssops (*Agastache* spp.)

9 stonecrops (*Sedum* spp.)

5 orange New Zealand sedges (*Carex testacea*)

PLANTING

compost

premium-quality garden soil

mulch

PREPARATION

These plants will cover a 3-metre (10-foot)-long by 3-metre (10-foot)-wide garden bed. If your garden bed is larger, then simply repeat the pattern with more plants.

Prepare your garden bed by turning the existing soil, then adding compost and additional premium-quality garden soil (see page 111).

METHOD

Tentatively lay out the Chinese silver grasses and honey bushes as per the blueprint to the right. Add the giant hyssops, stonecrops and orange New Zealand sedges.

When you're happy with the form of the layout, you can plant your garden bed. Add a layer of mulch approximately 5–10 centimetres (2–4 inches) thick.

Finish by watering in your plants.

WATERING

In the warmer months, water once or twice a week. If a heatwave is expected, then increase watering as required.

In the cooler months, water as required.

CULTIVATION

Chinese silver grass, giant hyssop, stonecrop: After the last frost, perform your winter cutback (see page 128).
Honey bush: If growth becomes too rampant, then perform a hard cutback to the ground in early spring.
Orange New Zealand sedge: Remove dead foliage by brushing with a hand rake in spring.

LIVING ROOM

ENTRY

GARAGE

WINDOW

PORTICO

FENCE LINE

① ② ③ ④ ⑤

DRIVEWAY

FENCE LINE

FOOTPATH

01 Chinese silver grass (*Miscanthus sinensis*)

02 honey bush (*Melianthus major*)

03 giant hyssop (*Agastache* sp.)

04 stonecrop (*Sedum* sp.)

05 orange New Zealand sedge (*Carex testacea*)

The perfect greenhouse

Many gardeners dream of having their own greenhouse. These warm, sheltered structures create the perfect microclimate to grow exotic species that wouldn't thrive outdoors. However, home greenhouses don't need to be used purely to house exotic specimens; they can allow you to plant produce earlier in the season so that your edible plants have had a head start before the weather picks up. I love mixing special tropical plants with my productive garden seedlings.

 Partial shade to full sun

KIT OF PARTS

PLANTS

3 orchids – choose your own preferred species

2 orchid cacti (*Epiphyllum* spp.)

1 tassel fern (*Huperzia* sp.)

2 mistletoe cacti (*Rhipsalis* spp.)

assorted produce or flower seeds

PLANTERS

3 medium terracotta planters, approx. 30 cm (12 in) diameter and 30 cm (12 in) high

2 medium hanging planters, approx. 20 cm (8 in) diameter and 20 cm (8 in) high

3 small hanging planters, approx. 14 cm (5½ in) diameter and 14 cm (5½ in) high

seed-raising trays or punnets as required

PLANTING

orchid bark

premium-quality potting mix

seed-raising mix

METHOD

Place a layer of orchid bark in a medium terracotta planter, approximately 5–15 centimetres (2–6 inches) deep. Place an orchid into the planter, and backfill with orchid bark. You can gently compact the bark, but it is important to have air pockets in the mix for the orchid roots to grow into.

Repeat for the remaining medium terracotta planters and orchids. Place the potted orchids on your growing bench so they're not on the cold floor.

Place a layer of potting mix into a medium hanging planter, approximately 5–10 centimetres (2–4 inches) deep, and place an orchid cactus. Backfill with potting mix, making sure to gently press down to compact the potting mix. Repeat for the other medium hanging planter and orchid cactus.

Place a layer of orchid bark into one of the small hanging planters, approximately 5–8 centimetres (2–3 inches) deep, and place the tassel fern. Backfill with more orchid bark.

Place a layer of potting mix into the remaining small hanging planters, approximately 5–8 centimetres (2–3 inches) deep, and place a mistletoe cactus into each one. Backfill with potting mix, making sure to gently press down to compact the potting mix.

Hang all of your hanging planters from the ceiling of your greenhouse so that the medium and small planters are layered in an organic way. Plant your produce or flower seeds into your seed-raising trays or punnets. Refer to the planting guide on your seed packets.

Finish by watering in your potted plants.

WATERING

In the warmer months, water once or twice a day. If a heatwave is expected, then increase watering as required.

In the cooler months, water as required.

IF YOU HAVE A GREENHOUSE, THEN YOU CAN OFTEN START GERMINATING YOUR SEEDS A LITTLE EARLIER THAN RECOMMENDED ON THE PACKETS. AS A GENERAL GUIDE, WAIT FOR THE WEATHER TO CHANGE SLIGHTLY TO THE PLANT'S REQUIREMENTS – THIS MAY BE AROUND TWO OR THREE WEEKS BEFORE THE RECOMMENDED PLANTING TIME.

CULTIVATION

Orchids: Remove spent flowers and dead foliage as needed.
Orchid cactus: Trim dead ends of foliage and remove spent flowers as they occur.
Tassel fern, mistletoe cactus: Remove dead leaves by using your fingers or pruning them with small snips.
Seeds: As they're tender during the germination stage, make sure to keep the soil moist but not drenched at all times.

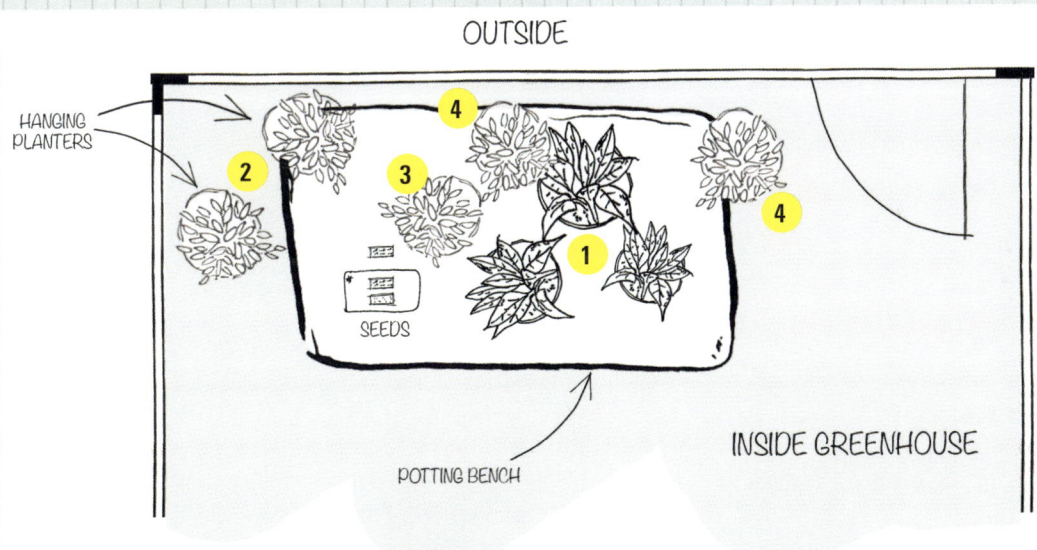

OUTSIDE

HANGING PLANTERS

2

4

3

1

4

SEEDS

POTTING BENCH

INSIDE GREENHOUSE

01 orchid – your own preferred species

02 orchid cactus (*Epiphyllum* sp.)

03 tassel fern (*Huperzia* sp.)

04 mistletoe cactus (*Rhipsalis* sp.)

Indoor forest

When I moved out of home, the place I lived in didn't have a garden big enough for my green thumbs – so I ended up filling my living room, kitchen and bedroom with plants. I figured that if I couldn't feed my desire to garden via my outdoor spaces, then indoor gardening was the answer. Creating an indoor forest is a simple way to nurture a special oasis that brings your garden inside. It doesn't have to be chaotic; rather, it should be curated in a way that makes your home tranquil.

 Filtered to bright light

KIT OF PARTS

PLANTS

2 kauri pines (*Agathis robusta*)

6 heartleaf philodendrons (*Philodendron hederaceum*)

3 fruit salad plants (*Monstera deliciosa*)

2 peace lilies (*Spathiphyllum* spp.)

PLANTERS

2 large grey planters, approx. 50 cm (20 in) diameter and 50 cm (20 in) high

3 short medium off-white planters, approx. 40 cm (16 in) diameter and 40 cm (16 in) high

2 tall medium off-white planters, approx. 40 cm (16 in) diameter and 50 cm (20 in) high

PLANTING

premium-quality potting mix

PREPARATION

Ensure that your planters have drip trays to prevent your floor from becoming stained.

METHOD

Place a layer of potting mix into the two large planters, approximately 10–20 centimetres (4–8 inches) deep, and place one kauri pine in each.

Backfill with potting mix, leaving enough room to plant three heartleaf philodendrons evenly around each kauri pine. Add the philodendrons, then continue to backfill with potting mix, making sure to gently press down to compact the potting mix.

Plant one fruit salad plant into each of the three short medium planters, then plant one peace lily into each of the tall medium planters.

Arrange the planters as per the blueprint to the right.

Finish by watering in your plants.

WATERING

In the warmer months, water once a week.

In the cooler months, water once a fortnight.

CULTIVATION

Kauri pine, heartleaf philodendron, peace lily: Trim away dead foliage and branches as they appear.
Fruit salad plant: Trim away dead foliage as it appears.

You can choose to plant a few heartleaf philodendrons around the peace lily as well as the kauri pine, and their foliage will drape over both planters.

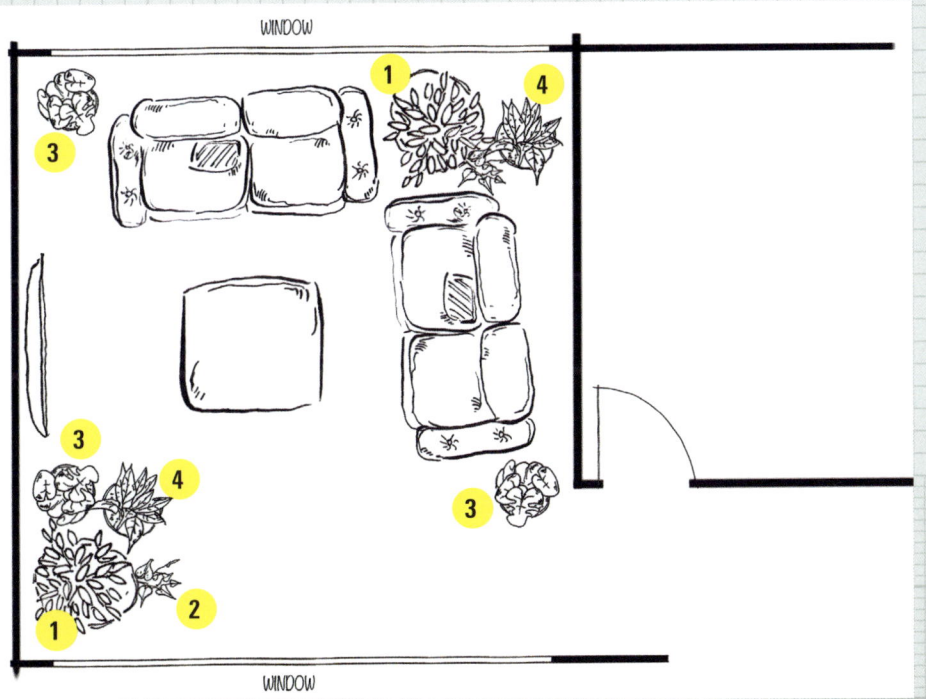

01
kauri pine
(*Agathis robusta*)

02
heartleaf philodendron
(*Philodendron hederaceum*)

03
fruit salad plant
(*Monstera deliciosa*)

04
peace lily
(*Spathiphyllum* sp.)

235

Flower-garden square

Flower gardens ooze joy. Thanks to their immense pops of colour and beauty, they're the ideal way to celebrate the seasons. They're mesmerising on a large scale; however, not all flower gardens need to roll across the hills. Even the smallest additions of flowers can lift the atmosphere in your garden. I'm a sucker for dahlias and roses, and teaming them with grasses can create a beautifully layered palette.

 Partial sun to full sun

KIT OF PARTS

PLANTS

6 coast tussock-grasses
   (*Poa poiformis*)

2 roses

3 dahlias

PLANTING

compost

premium-quality garden soil

mulch

PREPARATION

These plants will cover a 1-metre (3-foot)-long by 1-metre (3-foot)-wide garden bed. If your garden bed is larger, then simply repeat the pattern with more plants.

Prepare your garden bed by turning the existing soil, then adding compost and additional premium-quality garden soil (see page 111).

METHOD

Tentatively lay out the coast tussock-grasses, then lay out the roses and dahlias as per the blueprint to the right. When you're happy with the form of the layout, you can plant your garden bed. Add a layer of mulch approximately 5–10 centimetres (2–4 inches) thick.

Finish by watering in your plants.

WATERING

In the warmer months, water one to three times a week. If a heatwave is expected, then increase watering as required.

In the cooler months, water as required.

CULTIVATION

Coast tussock-grass: Tidy a messy-looking plant by removing dead foliage with a rake. Simply brush it through the foliage like a comb.
Rose: Deadhead the flowers once they're spent. In winter, prune stems back to promote new growth in spring.
Dahlia: Deadhead the flowers once they're spent. Wait until the top growth of your dahlias dies back or is killed by frost before you dig up your dahlia tubers. Store them in damp but not wet potting mix with sawdust, in a cool dark spot with good air circulation, during the cooler months, then plant them back in the garden bed once there is no more threat of frost.

STICK TO A SERIES
OF SIMILAR COLOUR TONES.
THIS AMPLIFIES YOUR CHOSEN
COLOUR AND CREATES A
COHESIVE DESIGN THREAD.
FOR EXAMPLE, IF YOU PREFER
RED, MAKE SURE ALL OF
YOUR ROSES AND DAHLIAS
HAVE THE SAME RED TONE.

SQUARE
TEMPLATE

01
coast tussock-grass
(*Poa poiformis*)

02
rose

03
dahlia

Grass pattern

Although they may seem simple, grass gardens provide complex habitats for a variety of wildlife. They're easy to maintain and provide a stunning landscape in the breeze. I love using a few species en masse to achieve a delicate aesthetic.

 Partial sun to full sun

KIT OF PARTS

PLANTS

3 giant needle grasses (*Stipa gigantea*)

5 *Calamagrostis × acutiflora* 'Karl Foerster'

5 beard grasses (*Andropogon* spp.)

5 yarrows (*Achillea* spp.)

PLANTING

compost

premium-quality garden soil

mulch

PREPARATION

These plants will cover a 2-metre (6-foot)-long by 2-metre (6-foot)-wide garden bed. If your garden bed is larger, then simply repeat the pattern with more plants.

Prepare your garden bed by turning the existing soil, then adding compost and additional premium-quality garden soil (see page 111).

METHOD

Tentatively lay out the giant needle grasses and *Calamagrostis × acutiflora* 'Karl Foerster' plants, then lay out the beard grasses and yarrows as per the blueprint to the right.

When you're happy with the form of the layout, you can plant your garden bed.

Add a layer of mulch approximately 5–10 centimetres (2–4 inches) thick.

Finish by watering in your plants.

WATERING

In the warmer months, water once or twice a week. If a heatwave is expected, then increase watering as required.

In the cooler months, water as required.

CULTIVATION

Giant needle grass, *Calamagrostis × acutiflora* 'Karl Foerster', beard grass, yarrow: After the last frost, cut back the plants to the ground. (For pruning advice, see pages 124–9.)

STYLING TIP
THERE ARE
MANY DIFFERENT
ORNAMENTAL GRASSES,
AND EACH ONE WILL
CHANGE THE LOOK
AND FEEL OF THE
PATTERN. WHY NOT
EXPERIMENT?

SQUARE
TEMPLATE

01
giant needle grass
(*Stipa gigantea*)

02
Calamagrostis ×
acutiflora
'Karl Foerster'

03
beard grass
(*Andropogon* sp.)

04
yarrow
(*Achillea* sp.)

239

Fernery grid

The delicate nature of ferns is an elegant addition to a sheltered garden under a tree. In the right location, ferns can be hardy. They come in a range of foliage textures and can be paired easily with an assortment of shade-loving species. When matching ferns with other plants, it's best to stick to foliage that is a solid shade of green with no variegation or other colours.

 Full shade to partial shade

KIT OF PARTS

PLANTS

3 cast iron plants (*Aspidistra elatior*)

5 blue star ferns (*Phlebodium aureum*)

3 bird's nest ferns (*Asplenium* spp.)

PLANTING

compost

premium-quality garden soil

mulch

PREPARATION

These plants will cover a 1-metre (3-foot)-long by 1-metre (3-foot)-wide garden bed. If your garden bed is larger, then simply repeat the pattern with more plants.

Prepare your garden bed by turning the existing soil, then adding compost and additional premium-quality garden soil (see page 111).

METHOD

Tentatively lay out the cast iron plants, then lay out the blue star ferns and bird's nest ferns as per the blueprint to the right.

When you're happy with the form of the layout, you can plant your garden bed.

Add a layer of mulch approximately 5–10 centimetres (2–4 inches) thick.

Finish by watering in your plants.

WATERING

In the warmer months, water once a day to three times a week. If a heatwave is expected, then increase watering as required.

In the cooler months, water as required.

CULTIVATION

Cast iron plant: Watch for scale and mealy bugs, and treat them as they appear. Trim any dead foliage as it occurs.
Blue star fern, bird's nest fern: Trim away dead foliage as it occurs.

This lush section of the fernery grid provides just a hint of the spectacular design when the plants are grown en masse.

→ SQUARE TEMPLATE

01
cast iron plant
(*Aspidistra elatior*)

02
blue star fern
(*Phlebodium aureum*)

03
bird's nest fern
(*Asplenium* sp.)

Simple woodland

There are beautiful examples of woodland gardens all over Europe and Japan. These gardens can claim the spaces between our buildings and add a sense of mystery to our urban environments. Once established, they grow effortlessly and offer a world of whimsy. A variety of textures, from flowers to bark, creates depth and an intriguing vista from your window. In narrow spaces, I like to play with verticality and use taller trees that have a narrow growth habit.

 Partial shade to full sun

KIT OF PARTS

PLANTS

5 silver birches (*Betula pendula*)

9 Japanese windflowers (*Eriocapitella hupehensis*) – white flowers

9 winter roses (*Helleborus orientalis*)

PLANTING

compost

premium-quality garden soil

mulch

PREPARATION

These plants will cover a 3-metre (10-foot)-long by 1-metre (3-foot)-wide garden bed. If your garden bed is larger, then simply repeat the pattern with more plants.

Prepare your garden bed by turning the existing soil, then adding compost and additional premium-quality garden soil (see page 111).

METHOD

Tentatively lay out the silver birches, then lay out the Japanese windflowers and winter roses as per the blueprint to the right.

When you're happy with the form of the layout, you can plant your garden bed.

Add a layer of mulch approximately 5–10 centimetres (2–4 inches) thick.

Finish by watering in your plants.

WATERING

In the warmer months, water two or three times a week. If a heatwave is expected, then increase watering as required.

In the cooler months, water as required.

CULTIVATION

Silver birch: Shape trees by removing untidy branches.
Japanese windflower: Deadhead the flowers once they're spent.
Winter rose: Trim spent flowers and dead foliage as required.

This straightforward three-plant pattern can be replicated as many times as necessary to create an appealing woodland strip in any narrow space.

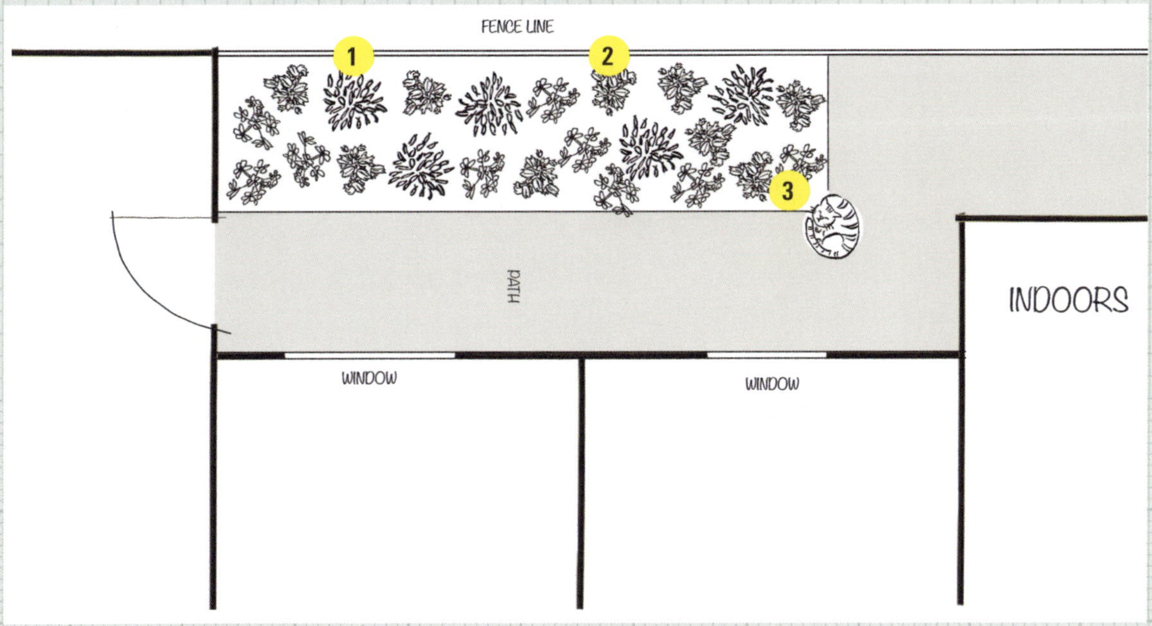

FENCE LINE

PATH

WINDOW

WINDOW

INDOORS

01
silver birch
(*Betula pendula*)

02
Japanese windflower
(*Eriocapitella hupehensis*)

03
winter rose
(*Helleborus orientalis*)

243

Thank you

As The Plant Society approaches its 8th anniversary, it's the perfect time to reminisce about our incredible journey – a venture that could happen only with the support of so many plant lovers, near and far.

To my partner, Nathan Smith, who forms the other half not only of me but also of The Plant Society. It's been an amazing adventure that I will always treasure and celebrate. What started as a simple idea has grown into what it is now because of you.

To our families, we are forever grateful. The support you've given us over the years has been monumental.

A huge thanks to the team at Murdoch Books for bringing my ideas to fruition. I am so grateful to have worked alongside a brilliant team, on a book that captures who I am now. Special thanks to Jane Willson for always hearing me out and believing in the concept, to Virginia Birch and Kristy Allen for seamlessly bringing it all together, and to Dannielle Viera for making sure that every word reads perfectly.

To Armelle Habib, I could not imagine anyone better to photograph this book. You are a legend.

From the get-go, I was in safe hands with Anna Collett. Working on this book has solidified how special you are and how fortunate I have been to have worked on three books with you.

To Trisha Garner, who beautifully captured my personality on paper, thank you for hearing me out and easing me each time I felt a bit stuck. To Joanna Hu, thank you for illustrating so many moments of joy.

Thank you to The Plant Society family for always working tirelessly to be our best. We are so proud of everything that we've achieved and look forward to growing together into the future.

Thanks to the myriad generous homeowners who welcomed us into their beautiful homes to capture their gardens. A special thanks to Heather Nette King, Ben Mooney, Kirrily Davis and Tony Nickels, Simone Haag, Eryca Green, Bridget Bodenham, Natasha Morgan, Patrick Roche and Peter Thannhauser, Anna Rozen and Taj Darvall, Nicola Rogers, Corina Baldwin and Rick Chazan.

Lastly, a huge thanks to all of the plant lovers and designers who have supported this journey. We are fortunate to do what we love because of your support.

Index

Published in 2024 by Murdoch Books,
an imprint of Allen & Unwin

Murdoch Books Australia
Cammeraygal Country
83 Alexander Street
Crows Nest NSW 2065
Phone: +61 (0)2 8425 0100
murdochbooks.com.au
info@murdochbooks.com.au

Murdoch Books UK
Ormond House
26–27 Boswell Street
London WC1N 3JZ
Phone: +44 (0) 20 8785 5995
murdochbooks.co.uk
info@murdochbooks.co.uk

For corporate orders and custom publishing,
contact our business development team at
salesenquiries@murdochbooks.com.au

Publisher: Jane Willson
Editorial manager: Virginia Birch
Design manager: Kristy Allen
Designer: Trisha Garner
Layout designer: Kristy Allen, Mika Tabata
Editor: Dannielle Viera
Writer: Anna Collett
Photographer: Armelle Habib
Illustrator: Joanna Hu
Production director: Lou Playfair

*Murdoch Books acknowledges the Traditional Owners of the
Country on which we live and work. We pay our respects to
all Aboriginal and Torres Strait Islander Elders, past and present.*

ISBN 978 1 92261 679 1

 A catalogue record for this
book is available from the
National Library of Australia

A catalogue record for this book is available from the
British Library

Colour reproduction by Splitting Image Colour Studio
Pty Ltd, Wantirna, Victoria
Printed by 1010 Printing International Limited, China

10 9 8 7 6 5 4 3 2 1